D0345439

Received On

FEB -- 2018

Magnolia Library

More Praise for EXPERIENCE ON DEMAND

"With sharp intelligence and lucid prose, Jeremy Bailenson deftly surveys the current state of this uniquely powerful medium. This engaging book shows us the vast potential of virtual reality as an empathy machine with the potential to change the world, while also being clear-eyed about its pitfalls and limitations."

— Laurene Powell Jobs, president of Emerson Collective

"What is virtual reality? Everyone is talking about it but very few have really indulged in artistic and well-crafted simulations. Jeremy Bailenson takes you beyond the hype and into the profound and empathic ways VR is enhancing every facet of life . . . from how we communicate to how we are entertained. A must read for anyone who is curious about the world we live in."

— Jane Rosenthal, producer and cofounder of Tribeca Enterprises

"Jeremy Bailenson's knowledge of VR, from high-level issues of hardware design and market dynamics down to details of human interaction and behavior is without equal, and this book is no exception. This book contains essential insights and information related to the human experience in virtual environments, and will be essential reading material for VR experience designers."

— Philip Rosedale, founder of High Fidelity and Second Life

EXPERIENCE
ON DEMAND

EXPERIENCE ON DEMAND

WHAT VIRTUAL REALITY IS, HOW IT WORKS, AND WHAT IT CAN DO

Jeremy Bailenson

W. W. NORTON & COMPANY

Independent Publishers Since 1923

NEW YORK | LONDON

For information about permission to reproduce selections from this book,
write to Permissions, W. W. Norton & Company, Inc.,
500 Fifth Avenue, New York, NY 10110

For information about special discounts for bulk purchases, please contact
W. W. Norton Special Sales at specialsales@wwnorton.com or 800-233-4830

Manufacturing by LSC Communications, Harrisonburg, VA
Book design by Chris Welch
Production manager: Anna Oler

ISBN: 978-0-393-25369-6

W. W. Norton & Company, Inc.
500 Fifth Avenue, New York, N.Y. 10110
www.wwnorton.com

W. W. Norton & Company Ltd.
15 Carlisle Street, London W1D 3BS

1 2 3 4 5 6 7 8 9 0

FOR CLIFFORD NASS,
THE KINDEST GENIUS I'VE EVER MET.

CONTENTS

EXPERIENCE ON DEMAND

INTRODUCTION

Mark Zuckerberg is about to walk the plank.

It's March 2014, and we're standing in the multisensory room of the Virtual Human Interactive Lab (VHIL) at Stanford University. I'm making last-minute adjustments to his head mounted display (HMD), the bulky, expensive, helmet-like device that is about to take him into another world. Zuckerberg, at the moment plunged into darkness, is asking me questions about the technical specs of the VR hardware in my lab—what's the resolution of the displays in front of his eyes? How quickly do the images on those displays update? Unsurprisingly, he is a curious and knowledgeable subject, and he's clearly done his homework. He's come today because he wants to experience state-of-the-art virtual reality, and I'm eager to talk to him because I have opinions on the ways in which virtual reality can be used on a social networking site like Facebook.

At Stanford, we are encouraged to face outward and to share our work, not just with academics but with decision makers of all types. I often do this kind of outreach, sharing my lab's unique capabilities with business executives, foreign dignitaries, journalists, celebrities, and others who are curious about the experience of virtual reality. On this day, I am eager to show Zuckerberg— someone who has demonstrated great philanthropic investment in areas of education, environment, and empathy research—the ways in which our work on VR has direct applications for those same issues. But first I have to show him what my lab can do. I usually start with "the plank"—it's one of the most effective ways to evoke the powerful sensation of presence that good VR produces. And our VR is *very* good—one of the best in the world. The floor shakes, we have "haptic" devices that give feedback to the hands, 24 speakers that deliver spatialized sound, and a high-resolution headset with LEDs on the side that allow cameras mounted around the room to track head and body movement as the user walks around the room. All of this information is assimilated to render interactive, digitally created places that allow the user to experience almost anything we can think of—flying through a city, swimming with sharks, inhabiting nonhuman bodies, or standing on the surface of Mars. Anything we decide to program can be rendered in a virtual environment.

The display comes on and Zuckerberg sees the multisensory room again, except that I and my lab assistants have disappeared. The room he is looking at is discernibly lower-res—a bit like television used to look before High Definition—but the carpeted floor, the doors, the wall in front of him are all there, creating an effective simulacrum of the space he was just standing in. Zuckerberg moves his head around to take it all in, everything smoothly scrolling into his vision as it would in real life. He steps forward and backward, and the illusion projected a few inches in front of his eyes corresponds with the movement of his body. "Trippy," he says. I lead him to a spot on the floor (I will be constantly at his side "spotting him" during this demonstration, as it's very easy to bump into real-world things when you're navigating a virtual space) and instruct my assistant in the control room to start the program. "Let's do the pit."

Zuckerberg hears an industrial whine, the floor shudders, and the small virtual platform on which he stands shoots away from the ground. From his perspective, which I can see via a projection screen on the wall, he's now standing on a small shelf about 30 feet in the air, connected by a narrow plank to another platform about 15 feet away. Zuckerberg's legs buckle a bit and his hand involuntarily goes to his heart. "OK, that's pretty scary." If we were measuring his stress signs we'd see that his heart rate was speeding up and his hands were beginning to sweat. He knows he is standing on the floor of a campus lab, but his dominant sense is telling him that he's precariously balanced at a deadly distance above the ground. He's getting a taste of "presence," that peculiar sense of "being there" unique to virtual reality.

Over the nearly two decades that I've been doing VR experiments and demonstrations, I've witnessed this scene—when a person is first enveloped by a virtual environment—thousands of times, and I've seen a lot of reactions. Some gasp. Some laugh with delight. Depending on what's being rendered in the program, I've also seen people cry out in fear, or throw their hands up to protect themselves as they hurtle toward a wall. An elderly federal judge once dove horizontally into a real table in order to catch an imaginary ledge after he "fell" off the virtual platform. At a demonstration at the Tribeca Film Festival, the rapper Q-Tip crawled across the plank on his hands and knees. Often my subjects just stand slack-jawed with wonder, gazing down, up, and around, amazed to see themselves suddenly surrounded by a digitally rendered world that nevertheless feels, in crucial ways, real.

It's a strange feeling, standing above the pit. No matter how prepared you think you are going to be the first time, it still takes you by surprise. Consider first that you, a subject immersed in the pit program, *know* you are participating in a VR demo—it's not like you accidentally wound up in a VR rig. So you're anticipating it. In fact, you've probably already witnessed—and perhaps even been amused by—another person going through the pit demo before you. You've looked at the projection of what they're seeing as they stand over the ledge. You've watched their legs bend as they lower their center of gravity, and seen their arms shoot out to their side for better balance. Watched as they tentatively shuffle across a thin beam that exists only in the program's code, and in the part of their brain that is interpreting the illusion projected into their eyes. Watched a person standing in the middle of the lab room, wearing an awkward helmet tethered to the ceiling by cables, ever so tentatively lean forward to gaze at a drop-off that doesn't exist.

In truth, it's kind of funny. And yet still, there you are a few moments later wearing the HMD yourself, your feet planted firmly on the same, solid floor you traversed just moments before, now suddenly facing a deadly drop with only a narrow plank to walk across. And it's not so funny anymore. If you're like one-third of the subjects in my lab, when I ask you to take a step off the platform into the virtual abyss, you'll refuse, standing rooted to a spot on the floor.

Zuckerberg *does* walk the plank, although it's not easy going. After he makes it to the other platform, I run a program in which his avatar grows a third arm and he has to learn how to move it by twisting his real limbs. Then he flies through the air like Superman. We put him in the body of a senior citizen and point him to a virtual mirror so he can see this strange doppelganger reflect the gestures he is making in the lab room. I load up another program, and he inhabits the body of a shark and swims around a coral reef.

"It's not bad, being a shark," he says. After a few more minutes he's had enough—the experience of VR can be overwhelming and even the best equipment today can cause eyestrain and head discomfort after twenty minutes or so.

For the rest of his two-hour visit to the lab we discuss my research into the psychology of virtual reality, and how it has convinced me that there are many ways the unique power of VR can be applied to make us better people, more empathetic, more aware of the fragility of the environment, and more productive at work. We talk about how VR is going to improve the quality and reach of education, and open up the world for people who can't afford to travel, transporting users to the tops of mountains, or into Earth's orbit, or into a calming oceanside setting at the end of a long day. How it's going to allow us to share experiences like these with our friends or family, even if they live far away.

"Experience" may seem like a generous description of what's happening when you're sitting in a chair or standing in a small space with goggles around your eyes, interacting with a digital environment. Experience is something that happens out in the "real world." It involves actually doing something. In our conventional understanding of the term, experience is "hard-won," it imparts wisdom, it's "the best teacher." We value it because we know that firsthand exposure to facts or events is the most powerful and effective way for us to learn and understand the world.

Sure, you might concede, media experiences can influence us, but they do so in much less powerful ways than actual experience. There are enormous differences between the physical world and the watered-down, abstracted versions of reality that we encounter even in multisensory media like film or videogames, and we can differentiate easily between those representations and reality. This is all true. But with virtual reality, the gap between "real" experience and mediated experience is about to get a whole lot smaller. The two won't quite be alike, but VR is far more psychologically powerful than any medium ever invented and is poised to dramatically transform our lives. It allows us to instantaneously conjure experiences at the click of a button. Experiences of any kind. One minute you could be sitting in a chair and the next you skydive, or visit an ancient Roman ruin, or stand at the bottom of the ocean. One day soon you will be able to share experiences like these remotely—with family and friends or people you just met who live on the other side of the globe.

VR will not only give us access to experiences that are difficult to obtain, it will also allow us to see impossible things, fantastic things, things that will allow us to see the real world in new ways and allow us to stretch our minds beyond anything we can imagine. You can become very small and peer inside a cell's nucleus or become immensely large and float in space and hold planets in your hand. You will be able to inhabit an avatar body with a different race or gender, or see the world from the perspective of an eagle or a shark.

There is a major qualitative difference between VR and watching video. It feels real. Good VR does that. When done right, VR experiences—intense, beautiful, violent, touching, erotic, educational, or whatever else you choose them to be—will feel so realistic and immersive they will have the potential, similar to experiences in the real world, to enact profound and lasting changes in us.

During Zuckerberg's visit that day we also talked about the pitfalls of VR. Like any transformative technology, it comes with significant risks. We discuss the dangers VR can pose to the physical and mental health of its users, and the detrimental effects certain types of virtual experiences can have on our culture as VR becomes a mainstream technology. I caution him about the current social costs of widespread addiction to seductive fantasy worlds, pornography, and videogames, and how those costs will be multiplied in a powerfully immersive medium. There's even the more mundane, but no less threatening, danger of millions of people banging into walls and coffee tables while blinded by headsets projecting compelling visions into their eyes.

A few weeks after my meeting with Zuckerberg, Facebook shocked the tech world by acquiring a tiny, crowdfunded company called Oculus VR for over $2 billion. At the time of the purchase, Oculus, founded by a 21-year-old self-taught engineer who had been mentored by the genius HMD maker Mark Bolas, had already reignited interest in VR among techies and gamers a few years earlier by making a lightweight and effective HMD prototype, the Oculus Rift, jury-rigged with smartphone screens and some clever programming. "I've seen a handful of technology demos in my life that made me feel like I was glimpsing into the future," wrote Chris Dixon, an investor at the influential Silicon Valley venture capital firm Andreesen-Horowitz. "Apple II, the Macintosh, Netscape, Google, the iPhone, and—most recently—the Oculus Rift."[1]

While the performance of this new consumer VR equipment was not quite as good as that of the state-of-the-art hardware in labs like mine, it was good enough to avoid the major performance problem that had bedeviled previous attempts at consumer VR—nausea-inducing lag. And, perhaps more important to the goal of making VR a viable consumer medium, the Rift could be manufactured for around $300—considerably cheaper than the $30,000 state-of-the-art HMD used at labs like VHIL. After a few false dawns, the long-anticipated arrival of affordable VR that actually worked had arrived.

Since the big Facebook acquisition in 2014, I have seen more innovation, growth, and excitement surrounding virtual reality than during the rest of my entire two-decade career. And it is accelerating. At the time of Zuckerberg's visit, virtual reality was an experience available only to the few people with access to research labs at universities, military facilities, hospitals, and corporations where VR was studied or used for a variety of applications, like training, industrial design, or medical treatment. Then, later in 2014, Google released its "cardboard" platform, which allows anyone with a recent model smartphone, for as little as $10, to turn their device into a VR headset. This gives users a limited, but still surprisingly compelling, VR experience at an incredibly low cost. Samsung has introduced a similar system called Gear, though one that costs a bit more as it has its own rotational tracking system housed in plastic. Whether entry-level systems like Cardboard and the Gear—which commonly deliver 360 video or a very limited level of immersion—technically qualify as VR is a matter of debate. For purists, VR requires motion tracking and digitally created environments that can be moved through. For the purposes of this book I will be defining VR broadly, to include the variety of immersive experiences on offer.

Thanksgiving 2016 was a strange holiday for me, as the traditional football programming on television was littered with advertisements for VR. Not just the Samsung Gear, which has been available to consumers for more than a year, but the second iteration of Google's VR system, called the Daydream, as well as Sony's PlayStation VR, which promises to transform gaming. Sony actually did a cross-marketing campaign with Taco Bell. I suspect this is the final sign that VR is officially mainstream.

At the higher end, more expensive VR systems (costing about $2,000, including the powerful computer required to run them), aimed at hard-core tech enthusiasts and videogamers, were just starting to be released. The HTC Vive and the long-awaited Oculus Rift are leading this charge. Unlike the more passive VR systems enabled by Cardboard and Gear, these higher-end systems are more immersive and are closer to the experience that can be produced in a lab like the one I run at Stanford. Combined with haptic devices, which provide touch and game controllers, they allow interactive engagement within digitally created worlds.

It is an exciting time to be working in VR. The sudden appearance of this new hardware is spurring an explosion of creative applications and content, as artists, filmmakers, journalists, and others try to figure out how the medium works. Investors are also bullish, with at least one tech-focused investor group predicting VR to be a mainstream technology, worth an estimated $60 billion, in the next decade.[2]

That doesn't mean that the VR rollout we'll see in the next couple of years is going to be completely smooth, or that there aren't still some considerable limitations to the technology. High-end VR is expensive, and headsets are still awkward to wear. Focusing on a screen a few inches in front of your face for an extended period of time causes eyestrain, and some people experience motion sickness during use. Room scale VR—the really immersive kind that lets someone walk around a scene—requires a dedicated empty room or at the very least a lot of empty space, which few of us are lucky enough to have in our homes. These are just a few of the hurdles VR designers are struggling with as they bring their devices to market. But considering the huge technical improvements that have been made in just the past few years, these challenges are surmountable.

Then there's the question of actually wearing the equipment. "Who is going to put on goggles?" some ask, pointing to the consumer failure of Google's much hyped augmented reality eyewear, Google Glass. Glass, of course, turned off a lot of people because it had the unnerving ability to seamlessly record video and audio. It was also considered antisocial, allowing people to seemingly interact with the real world while checking their e-mail. VR does not aspire to be integrated into one's day-to-day existence. For the near future at least, VR headsets will sit next to one's computer or gaming system, to be put on to experience a discrete piece of VR content or to socialize with others in a virtual setting. Perhaps an article one is reading online will have VR content attached, your brother will send a VR video of your nephew's graduation ceremony, or you'll decide you want to watch highlights from the NBA finals as if you were sitting in a courtside seat—you'll just put on the VR headset for 15 minutes or so. It's true that the idea of using the Internet with goggles on seems outlandish today, but then, a few years ago, so did a world in which everyone was staring at iPhone screens, or Skyping, or walking around city streets wearing massive, noise-canceling headphones. Once people get a taste of the experiences VR can bring them, the strangeness of the HMD will go away.

And what all this means is that intense virtual experiences will, sooner than many expect, be available to a massive consumer audience. Having studied VR for a few decades, I can tell you that this is no small matter. VR is not some augmentation of a previously existing medium, like adding 3D to movies, or color to television. It's an entirely new medium, with its own unique characteristics and psychological effects, and it will utterly change how we interact with the (real) world around us, and with other people.

And yet even as all this new content and technology is released in the next few years, few still understand how the technology works, how it affects the brain, and what it's useful for. This is why I'm writing this book.

This book is not meant to catch readers up on the latest trends in VR technology—that would be a fool's errand. Things are moving too quickly for that. But we are at a moment in time where it's a good idea to take stock of what VR can do—and what we want it to do. So instead, I've tried to focus on the bigger issues that are raised by a technology that allows us to inhabit and interact with virtual worlds in a way that feels more compelling than any other technology previously devised, with a particular emphasis on the positive applications.

Of course, trying to predict how a developing technology is going to affect culture is a speculative endeavor, at best. I was reminded of this while giving a talk at a technology conference in 2016 alongside Steve Wozniak, cofounder of Apple. Woz is high on VR—his first HTC Vive experience gave him goose bumps. But he cautions about overspecifying use cases. He told the story of the early days at Apple, and how when he and Steve Jobs made the Apple II, they conceived of it as a home appliance for computer enthusiasts, and believed users would use it to play games, or to store and access recipes in the kitchen. But it turned out it was good for unexpected applications. Sales really took off when a spreadsheet program was developed and suddenly people could do office work from home. According to Wozniak, he and Jobs were wrong about exactly what the Apple II would be used for. They knew they'd created something revolutionary, but they were mistaken in what that revolution meant. VR will be the same way. Almost everyone, upon trying it for the first time, can feel the significance and sheer magnitude of the technology. And yet we're still trying to figure out what it does best. My 91-year-old grandfather summed up the challenge well. After years of working with VR, I finally convinced him to put on the Oculus Rift and do a few demos. After a few minutes, only moderately impressed, he took it off, shrugged, and said, "What am I supposed to do in here?" He didn't say it in a pejorative way, but was clearly trying to understand the point of this amazing technology.

Consumer VR is coming like a freight train. It may take two years, it may take ten, but mass adoption of affordable and powerful VR technology, combined with vigorous investment in content, is going to unleash a torrent of applications that will touch every aspect of our lives. The powerful effects that researchers, doctors, industrial designers, pilots, and many others have known about for decades are about to become tools for artists, game designers, filmmakers, journalists, and eventually regular users, empowered by software to design and create their own custom experiences. At the moment, however, VR is unregulated and poorly understood. Consequently, the most psychologically powerful medium in history is getting an alpha test on-the-fly, not in an academic lab but in living rooms across the globe.

We each have a role in defining how this technology is shaped and developed. In this book I want to encourage readers to take a broader look at VR's applications—to look past the immediate offerings of games and movies and consider the wide array of life-altering things it can do. I will help readers understand it as a medium, and will describe some of the powerful effects I've observed in my nearly two-decades-long study of VR. This is so that, as we move from these toddler stages, we are using it responsibly, and making and choosing the best possible experiences for us—experiences that can change us and the world for the better. And the best way to start using VR responsibly is to understand what we're dealing with.

This is a unique moment in our media history, as the potent and relatively young technology of VR migrates from industrial and research laboratories to living rooms across the world. Even as we are amazed by the incredible things VR will allow us to do, the inevitable widespread adoption of VR poses unique opportunities and dangers. What do we need to understand about this new technology? What are the best ways to use it? What are its psychological effects? What ethical considerations should guide its use—and what practical ones, for that matter? How will VR change the way we learn, the way we play, or the way we communicate with other people? How will VR change how we think about ourselves?

What, when given a limitless choice, do we actually want to experience?

CHAPTER 1

PRACTICE MADE PERFECT

It was just one of dozens of snaps Stanford's quarterback Kevin Hogan would take that day against the Maryland Terrapins at the 2014 Foster Farms Bowl. The coaches had called for a simple running play, "95 Bama," which required Stanford's wide receiver to block one of Maryland's safeties after Hogan handed the ball off to his running back. But as the teams lined up, Hogan noticed a subtle change happening in the Terps' defensive formation. Their safeties began shifting positions, making it impossible for that crucial block to be made. Hogan realized if he didn't change the play in the seconds before the snap, the unblocked Maryland safety would stop the running play cold, leaving Stanford with a loss. So, exercising the kind of on-field decision-making that is crucial for a quarterback's success, Hogan killed the original plan and changed to a different running play, which would allow Remound Wright, his running back, to exploit the new space in the Terps' defense.

The call resulted in a 35-yard gain, and was just one of hundreds of small decisions Hogan made that day to help Stanford to a convincing Bowl victory. Hogan was later asked how, in that split second, he recognized the opportunity to change the play.

"It was easy," Hogan told the journalist. He was familiar with the Terps' blitz. He'd already seen it countless times with his own eyes, through a virtual reality training program adopted by Stanford earlier that season.[1]

When you picture big-time college or professional football, images of brutal physicality and stunning athletic feats come to mind. And for a few intense hours on any given weekend, that's what football fans get—crunching tackles, balletic catches, deft touchdown passes, and other displays of sublime athleticism. This is what fans see on ESPN and in YouTube highlight clips, and because of this focus on the extreme physical aspects of the game, it is easy for casual fans to forget how cerebral football at the highest levels is, not just for the coaches, but for the players as well. This focus on the mental aspect of the sport is reflected in the way teams train for games. Unlike athletes in most other team sports, who prepare by doing drills or scrimmaging, the reality of preparation for a football player is often more prosaic: much of the time is spent poring over playbooks and watching game film in order to learn the extensive, customized offenses designed by modern coaching staffs.

In football circles, the process of learning all these plays is called "installation," as if the players were human computers uploading a new operating system. But people are not computers, and the act of learning an offense is not done passively. It requires hours of rigorous, focused study. In the morning before practice, and in the evenings before bed, Monday through Saturday, from summer to winter. Over and over and over again. There's no other way to memorize these intricate game plans, so that, when it's time for kickoff, they have become so well ingrained that a player can carry them out unconsciously. Getting these plays right, and executing them effectively, is vital to a team's success. It is no surprise that in the big business of college and professional football, franchises spend a lot of time and money developing systems to improve this process. And no player is more responsible for implementing these systems than the quarterback.

Consider NFL veteran Carson Palmer. During the typical week of a season, the Arizona Cardinals' quarterback and his coaching staff will whittle an initial playbook of 250 unique plays down to about 170.[2] Each has to be studied and memorized. This includes not just the basic formations, positions, and movements of his teammates, but all the related information: What is the defense of this particular opponent likely to do? What should Palmer's response be if the defense changes its formation? If it's a passing play, which receiver gets the first look from the quarterback, and which one is the last resort? These various contingencies, too, must be learned, and they must be learned for *each* play. It's a dizzying amount of information, and in order to take it all in by game time Sunday, Palmer has adopted a disciplined regimen of almost continuous study. Each week during the season, Palmer and other top-level quarterbacks are like students with only a week to cram for their finals—a test which will be broadcast to tens of millions of viewers and dissected mercilessly on ESPN and sports radio the next day.

Palmer's week of intensive study usually starts on Tuesday night, after the coaching staff sends over the playbook for the following Sunday's (or in some cases, Monday's) game. During practice on Wednesday through Friday, the team will run through these plays. Traditionally, these practice plays will be captured on video and digitally catalogued, so players can review them on their computers or tablets off the field until they know them cold. But since the 2015–2016 season, the Arizona Cardinals have added the same VR technology that Kevin Hogan used at Stanford in 2014. On mornings and evenings before practice, Palmer slides on a headset and reviews practice footage captured with a 360-degree camera that has been mounted behind him on the practice field. When Palmer puts on the Head Mounted Display in his home office, he is instantly transported back to the moment in practice that he is reviewing, enveloped by an immersive re-creation of the play in which his real-life perspective of the unfolding action is reproduced almost exactly as he sees it on the field. For Palmer, and a growing number of professional, college, and even high school quarterbacks, virtual reality has literally been a game changer.

Palmer has seen a lot of different kinds of technology in his football career. When he was in high school in the late 1990s, he was still using a classic playbook, a three-ringed binder filled with hundreds of pages of Xs and Os diagramming formations. By then, film was ubiquitous: practice and game footage would be captured on videotape, filmed from a camera high up in a stadium press box for later study and review. As his career progressed, these basic technologies didn't change much. The quality of the video footage got better, and the number of cameras increased. Television broadcasts of his college games at the University of Southern California provided multiple cameras, which made it possible to review plays from different angles closer to the field. This also introduced some problems: the accumulating amount of footage, still captured in analog format, made finding a particular play to review a laborious process. Palmer still recalls with frustration the days of analog Betacam tapes. "If you wanted to find first down, or red zone, or first and ten," he told me, "you would have to go through these stacks . . . there was no categorizing it digitally like there is now. Now you just type in what you want and 'boom,' there it is. That was a huge, huge jump."

I spoke with Palmer at the end of the Cardinals' June 2016 minicamp. It was a few months after the most successful season in his career, during which he led Arizona to the NFC Championship game and the best record in franchise history. As a scientist studying virtual reality and a cofounder of STRIVR, the company that designed the VR system he was using, I wanted to know more about his experience with virtual reality, and why he felt it made him a better player. I'd gathered he was enthusiastic from the comments he'd made in a couple of articles published during the previous season that had become particularly popular around the STRIVR offices.[3] "It's blown me away," he told a writer for ESPN. "Literally six days a week I use it. . . . It's been a big part of my prep every week."[4]

I asked Palmer how VR compared with the other types of technology he'd used in his career. He described previous technologies—playbooks, tablet computers, even game film—as "prehistoric." "It's so much more beneficial than looking at film of someone else do it," he told me, "or a diagram or an overhead projector. . . . It's definitely helped in my preparation. . . . helped me absorb very complex systems, faster. I definitely got more reps."

"Experience," he added, "goes a long, long way."

HOW VR WORKS

The physical world changes when we move. Walk closer to a tree, and it gets bigger. Turn your ear toward a TV set, and it gets louder.

Touch your finger to a wall, and your finger feels resistance when it makes contact. For every physical action, there is an appropriate update to our senses. This is how humans have avoided bears, found mates, and navigated the world for millennia.

When virtual reality works well, it is seamless, and the virtual world changes just as the physical world does. There are no interfaces, no gadgets, no pixels. One second you are strapping on an HMD and the next you are somewhere else. That sensation of "being there," wherever the program you are running takes you, is what researchers call *psychological presence*, and it is the fundamental characteristic of VR. When it happens, your motor and perceptual systems interact with the virtual world in a manner similar to how they do in the physical world. Carson Palmer internalizes his playbook faster in VR than with video because of presence. Presence is the sine qua non of VR.

Let me give you an example of presence in action. In 2015, we did a shoot for a major network news program in my lab. The anchor, one of the regular hosts, put on the HMD and did about a dozen demos. All the while his crew filmed him from three separate camera angles. The demo that stands out from that day-long shoot was one we call "The Earthquake." In it, the user stands on a virtual factory floor, surrounded by heavy wooden crates stacked up to the ceiling. Each one is about the size of a desk, and they are stacked—rather haphazardly and precariously—about 10 feet in front of you and behind you.

For those of you who have been in a major earthquake, you know this is bad news. The good news is that in this virtual factory there is a very sturdy steel table to your left, tall enough for the user to fit underneath. It's the classic instantiation of "Drop and Cover," and we built this demo for the chief of the San Mateo County Fire Department with the idea of saving lives by teaching the muscle memory to survive an earthquake. Think of it as an earthquake survival simulator.

The network anchor put on the headset and looked around.

"Have you ever been in a quake before?" I asked.

He replied no, and I made sure he saw the table. "That is how you will save your life."

I then hit the "Q" button on the keyboard, initiating the quake in the program. The floor of our lab, which is made of a very rigid metal and designed to carry vibrations, began to shake and bounce. A thunderous rumble blasted through the lab's spatialized surround sound speaker system. We could see everything he was seeing through a monitor on the wall of the lab. In the virtual factory, the boxes started to sway and tilt, and it was very clear that the whole stack was coming down right on top of the anchor.

Few people can hide their reactions to this highly persuasive simulation—the heart rates of most people will speed up. Their hands will sweat. But for some people, the illusion is so powerful that their limbic system goes into overdrive. We call these people "high presence," and for them, VR is an especially powerful medium.

This anchor clearly fit that description. The simulation to him was psychologically real. He did exactly what we were trying to teach—he dropped to his knees and dove under the virtual table, put his head on the floor and his hands over his head. He reacted appropriately—took the actions to save his life. And he was clearly visibly upset by the quake.

Then something unusual happened. In our simulation, the boxes are stacked in the same starting point for every demo, but we model physics such that the effects are probabilistic. In other words, for every different earthquake—and we have run thousands of people through this—there is a different pattern of falling boxes. Sometimes they fall backward, sometimes forward, and the collisions and ricochets are unique every time. The anchor had an experience that I'd never seen before. I guess you could say he hit the jackpot. One of the boxes had the perfect trajectory and somehow bounced right under the table with him. There are only inches of clearance, but somehow it happened, and in his safe zone under the table one of the boxes came crashing into him.

He screamed, pounced to his feet and sprinted. In his virtual scene, he was sprinting to safety. In the physical world, he was running right at a wall. I managed to stop him before he collided, but it was a close one. At the outset of the experience, he had consciously known the simulation wasn't real, but in the moment, the illusion of presence took over. His brain reacted as if the box were a real danger. As far as his brain was concerned, the fake falling boxes could hurt him.

Matthew Lombard, a professor at Temple University who has been studying VR since the 1990s, defines presence as "the illusion of non-mediation."[5] On the tech side, we labor intensely to increase tracking accuracy, reduce the latency of the system, and do all the other magic it takes to build great VR. But for the user it's just a crate careening at your head while you are cowered under a table.

TRACKING, RENDERING, DISPLAY

Before we go further, you need to understand a few technical matters. In order to create presence, three technical elements have to be executed flawlessly: *tracking*, *rendering*, and *display*. One of the reasons consumer VR is possible at all is that the computing requirements to make these vital elements possible are only now inexpensive enough to sell to a large customer base. If any one of these elements is off, users can experience simulator sickness—

an unpleasant feeling that occurs when there is a lag between what your body tells you you should be experiencing, and what you actually see.

Tracking is the process of measuring body movements. In the earthquake demo, we were tracking body position in X, Y, Z space, and the subject's head rotation. In other words, if he walked forward a step (positive on Z), we measured that displacement in his body position. If he looked left (negative on yaw), we measured this rotation. We recently published what is called a meta-analysis—a study that combines the summary data from every paper we could find that has ever been published (and many that haven't) in an area. The meta-analysis was designed to understand the relationship to all of the features that make VR special—the affordances—and psychological presence. We wanted to understand what the relative benefits of technological immersion were on psychological engagement. We looked at about a dozen features, ranging from image resolution to display field of view to sound quality. Tracking was at the top of the list, and it ranked second among all of the cues with an effect size of 0.41, considered to be a "medium" effect by statisticians. Basically, this means that for every unit increase on the technology side of tracking there was a bigger increase in psychological presence compared to the other technological improvements.[6] In my lab, we take great pains to track accurately, quickly to avoid latency, and often at a high update rate. At talks, when describing virtual reality technology I often tell a joke. What are the five most important aspects of VR technology? The punch line: Tracking, Tracking, Tracking, Tracking, and Tracking. (I'll be here all week.)

Rendering is taking a 3D model, which is symbolic, mathematical information, and instantiating the proper sights, sounds, touch, and sometimes smell for the newly tracked location. When you look down at this book, you only see it from a very specific angle and distance. If you turn your head slightly, that angle and distance changes. In VR, every time a movement is tracked, the digital information in the scene needs to be rendered appropriately for the new location. It's not possible to store every possible viewpoint in a complex scene, so instead the viewpoints are rendered on-the-fly. The news anchor, when he dove to the ground, saw a new version of the room at every frame—which in our system in 2015 was 75 times per second. At each frame, we knew his precise location, and drew the floor closer and closer to his head as he dove down. We also rendered the sounds to be louder as he moved closer to the floor, since that is where the rumble was coming from. In VR, just as in the physical world, the senses need to be updated seamlessly based on movement.

Display is the manner in which we replace the physical senses with digital information. Once we render the sights and sounds for the newly tracked location, they need to be delivered to the user. For sight, we use a headset that can show stereoscopic information. As this book comes out, the typical headsets show images that are about 1,200 by 1,000 pixels in each eye, and update at 90 frames per second. For sound, sometimes we use earphones, while sometimes we use external speakers to spatialize sound. For touch, the floor shakes, and sometimes we use so-called haptic devices (more on this later).

The adoption of VR technology by athletes is only the latest chapter in a long and varied history of using virtual systems for training. In 1929, Edwin Link, an American inventor and aviation enthusiast, created the Link trainer. In his patent application (one of nearly 30 Link would receive in his lifetime), the machine was described as "a fuselage-like device with a cockpit and controls that produced the motions and sensations of flying."[7] We now know it as the original flight simulator, and many see it as an early example of virtual reality. According to Link's biography, it was inspired by the frustration he experienced when he took his first flying lesson. That lesson, in 1920, cost him $50 (over $600 today), and the instructor wouldn't even let him touch the controls. One can see it from the instructor's perspective—planes are expensive, and lives even more so. At the same time, we learn by doing, and Link was understandably eager to get his hands on the controls. It raised a vexing question: How could you teach someone a dangerous skill without putting them, and others, at risk?

In this problem, Link saw a business opportunity. It was the '20s, and the United States was in the middle of a civil aviation craze, creating a great demand for flight instruction. To remove the deadly risk of having beginners take the controls of an airplane, he created a fuselage that would move in three dimensions powered by pneumatic bellows, providing feedback to the student who was moving the controls. It was a great success—so much so that the military acquired his company in 1934, and by the end of the decade the trainer had spread to 35 countries and had been used to train untold numbers of pilots. In 1958, Link estimated that two million pilots had been taught on the Link trainer, among them half a million military pilots during World War II.[8]

Innovations in audiovisual reproduction and computers brought about digital virtual reality technology in the '60s, leading, in the decades that followed, to a host of virtual simulators for training specialists with difficult jobs, such as astronauts, soldiers, and surgeons. VR was adopted by these fields for the same reason that flight simulators became invaluable for training pilots. Mistakes are free in VR, and when the risks of on-the-job learning were high, technology that prepared a pilot, or a surgeon, or a soldier for the life-and-death responsibilities of his profession without risk was a huge win.

The use of VR for training would continue to grow. In the late '80s and '90s, early VR pioneers like Skip Rizzo at the University of Southern California began work with VR to help with physical rehabilitation for people who had suffered from strokes and traumatic brain injuries, and to train people to use prosthetics. These and other systems were designed to motivate users and ease the tedium of repetitious rehab exercises through interactive experiences. Some even provided feedback based on patients' movements, which reduced the potential for error when performing exercises. Studies showed these experimental therapies to be extremely effective. While there was a wealth of research showing that VR training was useful in a variety of domains, in 2005 I realized there were very few studies that specifically compared VR to other training techniques. Given how expensive it is to set up a VR training system, many from industry wanted to know exactly what their money would be paying for in terms of a gain in training. I decided to take a closer look, and my colleagues and I ran a comparative study to see how VR matched up against the most popular medium for training: video.

It's easy now to forget how much audiovisual technology has revolutionized education, but imagine you lived in a time before film was invented, and you were trying to learn how to dance, or swing a tennis racket, or perform even the simplest physical action without someone to teach you. Instead you would have nothing to go on but a diagram, or perhaps written or verbal instructions. Anyone who has tried to perform a repair on their car from a manual will immediately understand the challenge. The advantages of filmed instruction would have been obvious at the invention of moving pictures. So it's no surprise that educational films are as old as the medium of film itself. The aptly named Educational Pictures, for instance, a studio founded in 1915 in Hollywood, exclusively produced instructional films for a few years before it discovered it could make more money with comedic shorts.[9] As the century went on, instructional films flourished, enjoying even greater popularity with the US government. With the advent of inexpensive and portable recording technology on video in the late '70s, there was an explosion in instructional video.

Today, ubiquitous digital recording technologies found in inexpensive cameras and phones, along with many Internet distribution channels like YouTube, have made it possible for hundreds of millions of enthusiastic amateurs all over the globe to learn how to do things like paint, swing a golf club, fix a leaky faucet, or play "Stairway to Heaven" on the guitar. Learning this way isn't superior to good personal instruction, of course. You can interact with a real-life teacher, who can provide personalized feedback and motivation. But video instruction is considerably cheaper than hiring a personal trainer or tutor, and is vastly more detailed than previous forms of self-education.

For over a century, moving pictures represented the best medium for instruction in physical activities. And now here was VR, which could, by leveraging the power of presence, create the sensation that a virtual instructor was in the room right next to you. I wondered if the unique properties of VR could take this kind of learning one step further. And if so, how extensive were the benefits?

The activity we decided to study was the martial art of tai chi. The training forms of tai chi involve complicated, precise movements in three-dimensional space, but they were slow enough to capture by using the tracking technology that existed at the time. In the study, we split the participants into two groups, each of whom learned three of the same tai chi moves from an instructor. In one condition, the instructor demonstrated the moves on video. The other group of participants watched the moves as they were performed by a 3D virtual instructor, whose form was projected stereoscopically onto a screen in front of the participants. (Because the experiment involved physical movement, we did not want the students learning from the virtual instructor to wear cumbersome HMDs.) After the lessons were complete, participants were asked to perform from memory the series of tai chi forms that they had learned, which were recorded and sent to two coders trained to evaluate tai chi moves. These experts graded the participants on their accuracy. The findings showed that the group in the VR condition performed with 25% greater accuracy than the video group.[10]

Even with the limitations of the rendering systems in 2005, we were able to demonstrate that immersion in VR improved the learning of physical movements over a two-dimensional video, and to quantify that gain. Our tai chi study showed great promise for virtual instruction in fields like choreography, work training, and physical therapy, among others, and convinced me that improvements in the technology would one day lead to VR training simulations that could tutor users in complicated athletic movements, providing feedback and interactive instruction. This was good to know, as I would periodically be queried about this when athletes and executives from professional sports teams came through on tours of the lab in the decade following this study. How, they would ask, can this technology be used for football? For basketball? For baseball? At the time, there were rudimentary VR training tools for golfing and pitching, but those were mostly demos not designed for expert performers. No one, as far as I know, had successfully brought VR technology to the world of professional athletics.

There were a few very good reasons this was the case. As I've mentioned, until about 2014, head mounted displays and the computers required to run VR were still too expensive and difficult to use outside of research labs like mine. But that wasn't the only difficulty. Just making a virtual environment was time-consuming. A fully immersive football simulation, for instance, would require every detail, from the cones on the field to the folds in a jersey to the reflection off of a helmet, to be built up from scratch, one-by-one on limited budgets. Professional sports teams might have had the funds to make this kind of investment, but it would be too expensive and risky to implement an unproven sport-specific training technology in the already tight practice schedules of a professional team.

There were other difficulties: Who would code the scenarios for the teams? How would the virtual practices be designed? What about the technical hurdles? Creating effective computer simulations of a complex and dynamic experience like a football play would be extremely difficult, and the technology to capture video footage to work with VR didn't really exist. Sure, the research showed that using VR to improve the performance of high-level athletes was possible, even likely in the not too distant future. But using VR for athletic training was not something we were likely to see anytime soon. Like so many other VR applications already in use at corporate labs and research institutions like mine, the costs to bring them to a wider market were just too great.

In retrospect, I was being overly cautious in my forecasts on when the long-awaited emergence of viable consumer VR would actually come about. After studying VR for years, I was certain that someday virtual reality technology was going to eventually become mainstream, and was going to have a revolutionary effect on the way we communicate and learn. Still, few of us working in the field anticipated how quickly this day would arrive. But a perfect storm of technological advancements, economic forces, and the bold actions of a few entrepreneurs suddenly made that future, which once seemed decades away, arrive in a few short years. Cell phone manufacturers brought down the prices for screens. Lenses got cheaper. Computers got faster. People like Andy Beall at Worldviz created motion-tracking technology and design platforms that made the creation of virtual reality environments easier. Engineers like Mark Bolas came up with creative ways to make affordable hardware. Then, in 2012, backed by a hugely successful crowdfunding campaign, Oculus would begin manufacturing a prototype for the first high-end HMD that could be sold to a large consumer market.

After Facebook bought Oculus in March of 2014 for over $2 billion, you could suddenly sense the growing realization in Silicon Valley that something real was finally happening in VR. By January 2015, our lab's state-of-the-art HMD, the one that cost more than some luxury cars, had been replaced by developer models of consumer HMDs like the Oculus Rift and the Vive. These were smaller and lighter, worked just as well, and cost 1/100 of what we had been using. Now hundreds of developers were starting to develop content for these devices. And it was around this time that I was reacquainted with a former student of mine named Derek Belch, who also saw in that moment an opportunity to combine his passion for sports with the entrepreneurial VR fever that was sweeping Silicon Valley.

I had first met Derek Belch in 2005, a few years before the tai chi study, when he was a student in my class "Virtual People." Derek was also a kicker for the Stanford football team. (He has a place in the school's football lore for kicking a game-winning extra point in

Stanford's upset win over number-one-ranked USC in 2007, when
we were 42-point underdogs.) As an athlete, Derek was naturally
curious about how VR could be used to improve on-field performance. I told him the same thing I had been telling others when
the subject came up—that the technology just wasn't there yet.
But that didn't stop us from brainstorming after class about how a
VR training regimen might be designed, once it became possible.

Derek returned to the lab in 2013 as a Master's student in the
Department of Communication focusing on VR, right as the VR
boom was beginning. With the rapid improvements in VR hardware and the renewed interest from Silicon Valley, we agreed it was
the right time to pursue the study of sports training. The technology, it seemed, was finally ready.

As the 2014 school year started, Derek and I would meet twice
a week to talk over his thesis and how the technology could be
best utilized for elite athletes. We imagined a practice simulator
for football—a way for players to see formations and run through
plays, so that they could learn the offense and refine their ability
to read the intentions and tendencies of opposing defenses. Getting these kinds of reps is vital for high-level athletes, but running
through plays for the quarterback's benefit takes up a lot of time,
requires the participation of many teammates, and always carries with it the risk of injury during practice. A VR trainer could
capture important plays that the player would want to review,
and allow him to reexperience the practice of the play as much as
he wanted.

One thing we immediately agreed on was that the immersive environment must be photorealistic. Most of the study environments we used in the lab were computer-generated, but this
approach wouldn't work for sports training. An athlete must be
especially attuned to the small details in the game. The slightest

movement of an opposing player, a subtle lean in one direction, can give important clues about his intention and the entire course of a play. Accomplished athletes are attuned to such movements, and it would be vital for the VR-immersed player to see them. While it was theoretically possible for details like that to be created via computer graphics, it was not practical, and would require the resources and budget of a Hollywood digital effects department to pull off. We definitely didn't have that. Finally, real footage would help create the sense of presence, the sensation of "being there" in the virtual space, that was crucial to the learning experience we were creating.

360 video is now a familiar technology. Today, most large companies use it in one form or another, and even the *New York Times* produces 360 video content regularly. But in 2014 it was a challenge to produce. Coordinating the position and timing of 6 GoPro cameras simultaneously was a complicated task, and mounting the system on a tripod that was appropriately sized to work on a football field was an endeavor full of pitfalls. But when it is done right, a 360 video allows a person to put on an HMD and seamlessly look around a scene—one that updates to every micro-movement of his head and has high image quality. It's an amazing tool for quickly creating a highly realistic experience.

That spring, Stanford's coach, David Shaw, gave us the green light to bring our rig onto his practice field, a coup given how valuable practice time is, and how loathe coaches are to have their tightly choreographed schedules interrupted. But eventually we were able to get onto the field, film some plays, and get the content to the point where we were ready for our make-or-break moment— the demonstration for Coach Shaw. I'll never forget that hot day in April when I made my way to Coach Shaw's office to give the decisive demo of the system. After a computer crash and some fiddling with a wonky laptop straining under the processing requirements, I finally got it working. Coach put on the HMD and looked around as I ran him through a few plays. After about 45 seconds he took off the goggles and said, "Yes. That's it." Coach has a calm demeanor. Mike Bloomgren, the offensive coordinator, then bounced in, and after seeing the plays was hooting and hollering, dropping to his stance and yelling out calls.

That was the day I knew we had something.

The system was implemented for the 2014 season. It started with the inevitable technological hiccups. By late fall a firm training regimen had settled into place. Coaches would scout the defensive tendencies of upcoming opponents, and then Derek would film these scenarios in practice. After the film was stitched together into 360 video, Kevin Hogan would be able to run through practice plays as often as he liked, studying for the game in a much more intense way than he would by watching film or looking through playbooks.

Not long after a consistent VR training regimen for Hogan was implemented (using the headset for about 12 minutes before games), something remarkable happened. It is impossible to isolate any one factor to account for what happened to the Cardinals' offense at the end of that season, or any change in performance on any team. There are too many factors that influence success—schedule strength, new teammates, the inevitable peaks and valleys of performance that affect all athletes. Consequently, our policy has been not to assign too much credit to VR or to portray it as a wonder machine. Nevertheless, the statistics from that inaugural season certainly got our attention. After VR, Hogan's passing completion numbers went from 64% to 76%, and the team's total offense improved from 24 points per game to 38 points per game during this same period. But the most incredible numbers involved the success rate of scoring within the "red zone," or the space between the 20-yard line and the goal line. Prior to incorporating VR into their preparation, the Cardinals scored 50% of the time they entered the red zone—a poor rate of success. In their last 27 trips to the red zone in 2014, that rate improved to 100%.[11] Was this a regression to the mean, or was the STRIVR system giving Hogan an edge in his ability to read the game and make quick decisions?

Coach Shaw noticed the difference in Hogan right away. "His decision-making was faster. Everything was quicker," Shaw later said. "He saw things happening and could make those decisions and anticipate the ball coming out of his hands. . . . I'm not saying there is a 1-to-1 correlation, but it was along those lines. . . . We got him to think a little bit quicker. I think this virtual reality immersion and going through these plays helped him."[12]

After the season ended Derek had a meeting with Coach Shaw to discuss his future with the team. Shaw urged him to develop the VR training program and start a company. "He said 'get out of here,'" Derek would recall. "'You're a year ahead of everyone else. Go start a company.'" Shaw then became an initial investor in the business that would become STRIVR. (I am also an investor and cofounder of the company.)

Armed with statistics from the Cardinals' season and some scientific studies on VR learning he'd discovered during his Master's research, Derek took the $50,000 in initial investment money and traveled the country looking for customers. His goal for the first year was to find one team. But by the start of the 2014–2015 football season, he had signed up ten college teams and six NFL teams to multiyear contracts, including the Arizona Cardinals. It was an astonishing beginning and a great challenge. Suddenly Derek and the fledgling company were faced with the task of implementing individualized training regimens designed for teams playing at the very highest levels of the game, with an extremely promising but still experimental technology. This involved scaling up the company quickly and providing on-location support for the teams that were trying to figure out how to capture 360 video and incorporate that VR footage into their game preparations. The time-consuming stitching process that combined camera views into one seamless image required additional staff. Derek also hired some employees versed in data analytics to help the company crunch the numbers to measure how the players were improving as they used VR.

As that first season went on, it quickly became apparent that some teams were implementing VR training more than others. A few franchises barely seemed to be using it at all—it was difficult to tell, as many players neglected to log the minutes they spent using the HMDs. But as the season went on, one player was quickly standing out as a super-user, showing a consistent and steady use of STRIVR footage in his game preparation. This was the Arizona Cardinals' quarterback Carson Palmer, who, having just come back from a season-ending knee injury the year before, was enjoying a career year.

Thirty-six years old and a seasoned veteran, Palmer didn't particularly care for the technological innovations that teams seemed to be experimenting with constantly. "I don't buy in to all the new technology," Palmer said in November of 2014. "I'm archaic. I thought, 'There is no way this can change the way I play quarterback. But I am all in on this."[13]

When the season came to an end, Palmer had led Arizona to the franchise's best record of 13–3 and put up career best numbers in passing yards, passing touchdowns, and quarterback rankings. The Cardinals would make it to the NFC Championship game.

"THE WHOLE, BIG PICTURE"

What goes through the mind of a quarterback during those frenetic seconds after the ball is snapped to him in a game? I get a sense of this when I put on the STRIVR HMD and watch some of the demo footage taken at a Stanford practice a few seasons ago. It's a view only people who have played or coached at the highest levels of the game ever get a chance to see. When the program begins, I'm on a turf practice field on a bright, blue-skied day. There are white clouds in the distance. As I look left and right I see a row of five offensive linemen arrayed in front of me. I can see 11 defensive players beyond them, some right up on the line of scrimmage, others running around trying to confuse me and disrupt my reads. Everyone seems enormous and extremely close. There is a cadence and then the ball is snapped. What I immediately notice are the five massive linemen lunging toward me from only a few feet away. There is also a flurry of activity in my peripheral vision. Because this is practice there is little contact, but I'm still amazed by the speed and power of everyone around me. I think I see a receiver running downfield to the left of my field of view. And then, almost instantaneously it seems, the play is over and the screen fades to black.

For those who haven't spent thousands of hours playing the sport, the sequence seems like chaos, everything moving too quickly to make sense of. My untrained eye is unable to distinguish between important and unimportant details. I see clouds, red jerseys, the linemen and their movements, and the receiver in the corner of my field of view, but I don't know how to make sense of any of it. But a quarterback sees the events much differently.

Later, I asked Palmer how he can react to so much happening so quickly. "You don't notice little things," he told me, "you notice entire things. The whole, big picture. Without really focusing on little things. You're using your peripheral (vision) on the entire picture." Experts in perceptual learning call this process "chunking." It's how all the disparate parts of a complicated cognitive activity become one. For instance, when you first learn how to ride a bike, you are all over the place. You are thinking about your arm and your leg, and as you learn through experience and trial and error, you begin to only need to focus on a few things. And as it goes more smoothly it becomes a big chunk. Your brain has become more efficient at this task. Eventually, you are riding a bike without thinking about it, allowing you to turn your attention to the many other things you need to think about on a bike—like other cyclists or cars or potholes. Excellence is about attention and the allocation of resources.

The reason expert-level players like Carson Palmer can so efficiently process all this information is that they have amassed—through practice, study, and gameplay—so much experience that they are able to create and call upon extremely refined "mental representations" of what's happening around them on the field. This concept of mental representation in part comes from the research of K. Anders Ericsson, who has spent his career studying expert performers in a variety of domains, from those who excel at memorizing long strings of numbers, to chess players, to professional athletes in sports from rock climbing to soccer. Chess players, for instance, have played so many games they know which parts of the board need their attention and which ones don't. They can look at a chessboard and in a matter of seconds know what the correct move is. An amateur player will waste a lot of energy visualizing possible moves that an expert can dismiss right away.

Ericsson's research shows that mental representations are honed by deliberate practice, a particularly engaged form of learning that is distinguished by a motivated learner with well-defined goals, who gets feedback from his performance and has ample opportunities for repetition. Palmer's method for preparation, which involves quizzing himself as he goes through the plays, satisfies all these conditions. But perhaps most importantly, VR gives him access to unlimited repetitions. As he told me, "There is no other way to learn but to get the reps, to get the experience, [VR]'s just as close as you can get to another rep." It should come as no surprise that, as Ericsson has pointed out (at a time before immersive video systems like STRIVR), the most successful quarterbacks are "generally the ones who spend the most time in the film room, watching and analyzing the plays of their own team and their opponents." What's happening today is that the film room is becoming an immersive virtual space that bears more resemblance to the actual practice field than it does to two-dimensional images within a frame.[14]

Another advantage of VR is that, because users' brains are treating the experience they are having as psychologically real, they are physiologically aroused in a way that is similar to what occurs during real experience. The sights and sounds of the practice field, along with the sight of large linemen charging, heighten the emotional state of the player using the trainer, which improves learning. You're not sitting in a bucket of ice after the game, looking at an iPad; you are in there. This, perhaps, explains why VR also works so well as a visualization tool. There's a lot of literature showing that visualizing actions helps with performance. The mere act of thinking about an action, for instance, produces brain activity similar to what you'd see if the action were actually being performed. But that is only if the visualization is done the right way. The problem, of course, is that there is a high variability in how well people can visualize and put themselves in those situations. With VR, coaches and trainers can create visualizations for players.

Yet another reason VR works extremely well for learning is that it utilizes body movements. When you experience a VR simulation, you move your body as if you were experiencing an actual event in the real world. What makes VR different from using a computer is that you move your body naturally, as opposed to using a mouse and a keyboard. Hence, learners can leverage what psychologists call embodied cognition.

Embodied cognition argues that, while of course the mind is located in the brain, there are other organs in the body that influence cognition. Muscle movements and other sensory experience help us understand the world around us. When we think, the parts of the brain that are associated with body movements become activated. Consider a 2005 study on dancers.[15] Scientists studied two groups of professional dancers—experts in ballet and capoeira. Dancers watched some recorded dance moves of both styles while their brain activity was recorded in fMRI. When the dancers saw the dance moves from their own area of expertise, the "mirror system" of their brains became activated, while when they saw the dance moves from the other style, their brains did not activate as strongly. In other words, when they saw the moves that they had performed thousands of times over their lifetime, their brains activated as though they were performing those actions. So simply watching and thinking about dancing caused the dancer's brain to activate as if she were doing the moves herself. The brain understands watching an event by visualizing motor movements.

This activation of the motor action part of the brain can actually predict learning. In a study published in 2008 in the *Proceedings of the National Academy of Sciences*, scholars from Carnegie Mellon University examined hockey players, hockey fans, and novices to hockey. Hockey experts were better at understanding hockey moves than the novices, and this difference was accounted for by brain activation. In other words, the more brain activation in the higher-level motor areas that are associated with expertise from the viewer, the better he was at understanding the hockey moves. While these data are correlational, proponents of embodied cognition posit that by simulating an action in the brain, one can improve learning. The authors of the hockey study conclude that "The impact of athletic experience on comprehension is explained by greater involvement of brain regions that participate in higher-level action selection for those with hockey playing and watching experience."[16]

The impact of sensorimotor simulation in the brain also applies to basic science learning. In a 2015 study on college students learning physics, some learners were given actual physical experience with torque and angular momentum by spinning bicycle wheels around, while others just watched the wheels spin. Students who experienced the forces associated with angular momentum did better on a subsequent quiz than students who only watched the spinning wheels. And similar to the hockey study, the increase in performance could be attributed to the activation of sensorimotor brain regions which were measured later on when students performed similar physics tasks. This study shows first that people learn better by doing than by watching, but also that those who learn best are simulating motor action in the brain.[17]

We are still in the very early stages of understanding how best to use VR to maximize performance, which makes the gains shown in the early data so impressive. Going forward, STRIVR and other systems like it will be able to collect and analyze the massive amounts of data gathered by players like Palmer, who has shown himself to be especially adept at teaching himself, and discover the best ways to implement the training.

THE FUTURE OF VR TRAINING

STRIVR's success in sports training eventually got the attention of the corporate world. It turns out that all the things that make VR useful for training a quarterback—quick assessment of a scenario, improvement of decision-making in chaotic situations, and of course the ability to practice when the stakes are virtual instead of real—are also surprisingly effective at training employees. Walmart, the world's largest retailer, signed up with STRIVR to build an app to train their employees. In a Herculean effort, one of our executives read Walmart's entire training manual cover to cover and picked out the simulations that would be most conducive to using VR.

The first module we built was in the supermarket. For the deli counter manager, this means practice on handling multiple customers at once, and making sure that when there is a long line you don't ignore people while handling the current customer. For floor managers, this means quickly walking down an aisle to see if the rolls of plastic bags hanging over the aisles are fully stocked, or noticing that one of the customers is lingering a bit too long in one place (is she shoplifting?). I learned that if you stack the corn cobs too high on the shelf in a supermarket, the ventilation gets blocked and it is a violation of code. It's an easy thing to program into VR. Spotting this error may not be as glamorous as picking up a hidden blitz and then throwing the winning touchdown pass in a playoff game, but it's the kind of small efficiency that makes a difference. We conducted a pilot study in 30 training academies to ensure that people would actually use and enjoy the system (they did). It turns out they also learned better, as Walmart saw the difference in measured performance, and decided to scale up to all 200 of their training facilities.[18] Walmart is building a library in which the written manual will come with a set of experiences to train on. The beauty from their perspective is that VR is magnitudes of order cheaper than setting up a physical training store that is stocked with food and customers. But price aside, it's also more consistent—every trainee gets exactly the same experience, on demand.

When you start to consider these varied and powerful ways that interactive and analytical elements can be embedded in immersive virtual environments, the possibilities for VR training become endless. Training for soldiers, pilots, drivers, surgeons, police officers, and other people who are doing dangerous jobs—these are just a few of the hundreds of applications being used. But there are countless ways VR can be used to improve the cognitive skills we use in everyday life. Negotiation, public speaking, carpentry, machine repair, dance, sports, musical instruction—almost any skill can conceivably be improved through virtual instruction. We will see these and applications not yet dreamed of in the coming years as the consumer market grows, the technology develops, and our understanding of how to best use VR effectively improves.

This is a truly exciting and revolutionary time for education. Already, the Internet and video technology have opened up new opportunities for learning, and VR is going to enhance this development. There is so much untapped and untrained potential in the world. We have been taught to think that high-level performers are imbued with natural gifts that lead inexorably to greatness. While it is true that certain people are gifted with extraordinary natural abilities, it is only when that potential is combined with hard work and the proper coaching that it thrives. How many individuals are not meeting their potential because they lack the access to good instruction and learning tools?

I wonder, sometimes, at how frequently expert performance in specialized skills is clustered within families. Take the Manning family in football, which boasts three top NFL quarterbacks in only two generations, with one of them, Peyton Manning, rated among the very best to have ever played the game. It is clear that the Manning family is specially gifted with some of the natural qualities required to become a great NFL quarterback. But does that tell the whole story? Isn't it also likely, especially knowing how much success at that position depends on decision-making based on experience, that brothers Eli and Peyton benefited from growing up with a father who was also a professional quarterback, who could explain the nuances of the game to them and instruct them on the fundamentals of playing it from a very early age? What about all the kids who share similar physical and mental gifts with the Mannings, but who never reach that potential because they haven't had access to expert coaching?

For me the most exciting aspect of VR instruction is its potential to democratize learning and training. To be sure, it won't be as easy as uploading a kung fu program in a few seconds, like Neo does in *The Matrix*. Learning expert skills takes dedication and focus and lots and lots of practice. But it does mean that eventually everyone will have access to resources that, should they be willing to put in the work, can put them on the path to expert performance. There is more pressure than ever before for high performers in all domains, and for specialization and focusing at younger ages. It's a stubborn fact that those with access to special tutoring and instruction have a great advantage over those who do not. Just as access to online videos and instructional courses have opened up opportunities for learning, so will VR. It will take time, of course, before the HMDs and the content are inexpensive enough to be adopted by those who lack access to the kind of specialized instruction that is necessary to succeed in our increasingly competitive world, but when one considers how quickly data phones and their many apps have become ubiquitous, that future may come sooner than we think.

Sometimes, we don't value experience enough. Imagine a world in which the best teachers in all fields, realized as interactive embodied agents, are ready to guide promising minds through the lessons and practice necessary to thrive. VR training could go a long way in opening up real opportunities to millions whose talents are not being used.

CHAPTER 2

YOU ARE WHAT YOU EAT

Students of film will be familiar with the apocryphal story of the Lumière brothers screening their film, *The Arrival of the Train at La Ciotat Station,* in Paris in 1895. According to this legend the audience screamed in terror as the filmed train projected on the wall seemed to rush at them. We like this story because it speaks to the lack of sophistication of early filmgoers. *We* would never fall for a trick like that nowadays, of course. We can tell what is real and what is fake. Once the novelty of a new medium wears off, we like to believe, we can handle whatever it throws at us.

While there is some truth to that, there is a world of difference between the kind of immediacy and reality that VR creates, and a flat (or even 3D) projection on the wall. Having watched thousands of people in my lab, many of them repeatedly experiencing VR, I can tell you that VR experiencers behave differently. It takes a lot of viewings and a lot of willpower not to flinch when the crates start falling in the earthquake simulator.

When VR is done right, all the cumbersome equipment—the goggles, the controller, the cables—vanishes. The user becomes engulfed in a virtual environment that simultaneously engages multiple senses, in ways similar to how we are accustomed to experience things in our daily "real" lives. This is distinctly different from other media experiences, which only capture fragmented aspects of what our senses can detect. For instance, the sounds you hear in good VR don't come from a speaker rooted in one place. Instead, they are spatialized, and they get louder or softer depending on the direction you are facing (or if you are in a tracked environment, how close you move to the source of the sounds). When you look at something in VR, it is not framed by the dimensions of a monitor, or television set, or movie screen. Instead, you see the virtual world as you see the real one. When you look to the left or right, the virtual world is still there.

VR technology is designed to make itself disappear. When we watch a video on a screen, we have constant reminders of the artificiality of the experience. Film media are distinguished not only by the two-dimensional flatness of the image, but by the framing of shots, the unnatural camera movements, the cuts and other editing tricks, and perhaps most importantly, the unusual perspectives created by the camera placement. These all violate the rules by which we normally encounter the world with our senses. In my lab, when I give a visitor a particularly scary or intense demo, sometimes I will give them the "veteran's secret"—if you close your eyes, it goes away. Almost nobody thinks to do this on their own, as the strategy wouldn't work in the real world (you would still actually fall in the pit into the real world regardless of whether your eyes were open or closed).

My point in mentioning all of this is that the illusion of presence is unusually powerful. Even a naïve Parisian at the end of the nineteenth century could look at a wall on which images were being projected and determine that the train hurtling toward him was a fake. But if I put an HMD on Steven Spielberg and put him on a virtual boat with the shark from *Jaws* heading straight at him, there's a reasonably good chance that he'd be terrified.

The power of presence is not good only for cheap thrills—as I will show throughout the book, the psychological effects of VR can be profound and long-lasting. Study after study has shown the experiences that people have in VR have an impact on them. Their behaviors can change, and these changes don't disappear right away. This leads to a conclusion that captures the considerable promise and perils of the medium. VR feels real, and its effects on us resemble the effects of real experiences. Consequently, a VR experience is often better understood *not as a media experience, but as an actual experience*, with the attendant results for our behavior.

VR is an experience generator. Because it is a digital medium, anything we can imagine seeing or hearing can be easily generated in a VR environment. (The other senses are more complicated.) This opens up some inspiring possibilities about the kinds of experiences we might want to have, and how those experiences can make us, and the world, a better place. On the other hand, if we choose to experience unhealthy environments and experiences, we can expect unhealthy results. As a colleague of mine puts it, media experiences are like your diet: you are what you eat.

If you've taken a Psych 101 class, you are probably familiar with Stanley Milgram's famous studies on obedience, which remain one of the most famous and disturbing analyses of human behavior ever conducted. It was the 1960s, and academics were trying to understand what went so horribly wrong with humanity only a generation earlier, particularly how so many people acquiesced and even willingly took part in the evils perpetrated by Nazi Germany. As you'll recall, Milgram's study was based upon a subject who administered a test to a confederate—a person who, unbeknownst to the participant, was actually an actor taking part in the experiment. If the confederate's answer was incorrect, the participant was to give a shock to the confederate, with the voltage increasing in 15-volt increments for each wrong answer. The participants were led to believe that for each wrong answer, the learner was receiving actual shocks. In reality, there were no shocks. After a few wrong answers, the confederate would bang on the wall in pain and complain about his heart condition. If the participant continued, eventually the confederate would stop responding, as if he had passed out or worse. All the while, there was an authority figure in a lab coat telling the participant that he must continue, saying things like, "The experiment requires you to continue," and "You have no other choice but to continue." As the experiments progressed, Milgram measured two critical variables: how many shocks the participant was willing to deliver, in spite of the clear ostensible pain from the confederate, and also the effect that the obedience had on the participant.[1]

The results of repeated studies demonstrated that a majority of experiment participants obeyed the commands of the authority figure to the very end of the study, including the final 450-volt shock (labeled "Danger: Extreme Shock," on the shock generator). As can be seen in some very powerful and troubling videos that are easy to find online, the participants who obeyed did not do so without cost—often they would sweat, bite their lips, exhibit nervous fits of laughter, or groan and tremble. When people talk about the Milgram experiments they often focus on the terrible insight that people are capable of inflicting great cruelty, blindly following orders. But what is not emphasized enough is how terribly participants suffered while doing it.

In 2006, Mel Slater, one of the pioneering researchers in Virtual Reality, decided to replicate this experiment in VR.[2] However, there was a twist: he had participants engage in a task similar to the one in the original Milgram study, but in his study the confederate was replaced by a digital representation of a person. Furthermore, he informed participants that this virtual human was an "agent," meaning it was completely controlled by a computer (unlike a digital representation controlled by a person, which is known as an "avatar"). The original Milgram study was designed to deceive subjects into believing an actual person was getting shocked, but Slater was completely forthcoming, and all participants knew they were simply "shocking" a computer program. Slater himself played the experimenter/authority figure, sitting beside the participants while they administered the test (and the shocks) to the virtual agent. However, unlike in the previous studies, Slater did not urge the participants to continue against their will; they were told clearly and repeatedly they could withdraw from the study any time without penalty. He instead was interested in measuring whether or not inflicting harm to a purely virtual character caused anxiety.

The results demonstrated that despite knowing the recipient of the shocks was a computer-generated fiction, people responded psychologically as if it were a real person, both in terms of their behavior and their physiological responses, which Slater measured using heart rate and skin conductance. Even though they were consciously aware that everything about the study was a simulation—the lab environment, the "shocks," the learner—the subjects' brains treated it, to a significant degree, as if it were really happening. "The voices of some participants showed increasing frustration at [the Learner's] wrong answers," wrote Slater and his colleagues in the study. "At times when the Learner vigorously objected, many turned to the experimenter sitting nearby and asked what they should do. The experimenter would say: 'Although you can stop whenever you want, it is best for the experiment that you continue, but you can stop whenever you want.'" Slater noted a variety of reactions as participants administered higher degrees of shocks. Some stopped before the end. Some giggled at the Learner's protests (which also happened during the original experiment). And some showed genuine concern. "When the Learner failed to answer at the 28th and 29th questions, one participant repeatedly called out to her 'Hello? Hello? . . .' in a concerned manner, then turned to the experimenter, and seemingly worried said: 'She's not answering.'"[3]

What to make of this strange result? It's one thing to instinctively react to a virtual pit opening up beneath your feet, but experiencing a moral discomfort for virtual behavior toward an animated virtual human? When we consider that the subjects were made uncomfortable by the idea of administering fake electric shocks, what can we expect people will feel when they are engaging in all sorts of fantasy violence and mayhem in virtual reality? This is just one of the many concerns I hear constantly from visitors to my lab.

My colleague in the Department of Psychology at Stanford, Benoit Monin, is an expert in the psychology of morality. My lab collaborated with him to examine the impact of witnessing immoral events in VR. In a sample of over 60 participants, about half of them experienced a moral event.[4] In VR, they saw a character of their same gender giving first aid kits to 60 others who walked toward him. After receiving the first aid kit, each character turned and walked away. The moral stimulus was designed to grow more intense over time; the character gave first aid kits to 20 soldiers first, followed by a mix of 20 women and children, and then a mix of 20 children and elderly individuals. The simulation lasted approximately 5 minutes. The other half of the subjects experienced the immoral condition. Instead of giving first aid, the character punched the same sequence of 60 virtual humans who walked toward him/her. After being punched, the characters fell into a nearby pile. The pile of virtual bodies grew quite large over the course of the manipulation. As horrible as that sounds, it was designed to simulate the events that occur in typical media such as films and videogames.

Previous research in psychology has examined moral cleansing. In one study published in *Science*, subjects who were prompted to think about instances of immorality were more likely to take antiseptic wipes than subjects prompted to think about instances of morality.[5] The authors deemed this the "Macbeth effect": that is, a threat to one's moral purity induces the need to cleanse oneself. In our study, after the subjects took off the HMD, we offered them waterless hand sanitizer and measured how many pumps they took. Those who witnessed the immoral event used more pumps on average than those who witnessed the moral event, providing preliminary support for the notion that they needed to cleanse after the horrible experience of watching women and children get beaten. It's important to note that this was a small pilot study, and the results were modest, with a statistically significant but small effect size. But it resonates with the findings from Slater's study—intense events in VR have psychological effects.

Throughout human history every new communication medium has introduced its share of anxiety about its potentially nefarious uses and deleterious effects on people. Readers interested in technology are no doubt familiar with some well-known historical examples. That Socrates feared literacy because he believed the written word would degrade people's memories. That novels were eroding the ability of readers in the nineteenth century to distinguish between fiction and reality. Often these facts are cited by those who dismiss claims about media effects as overblown, as examples of how ridiculous and exaggerated people's concerns are about, say, violence in videogames, or the influence of digital culture on our ability to think. I think one can concede that sometimes these fears are exaggerated—but still take media seriously. Literacy and books *have* changed the way we think. Media images *can* have powerful effects on our minds.

I teach the class at Stanford called "The Effects of Mass Media," and the extent of these effects is what I study. I'm generally unalarmed about the extent to which traditional media affect us. But as absorbing as books or videogames or television are, in their immersive effects they pale next to virtual reality. Other forms of electronic media simulate fragments of the sensory input out of which we construct our conscious experience of reality. For instance, film or television or a video watched on a tablet may convey sounds and sights captured from the "real world," but when we interact with these media we are almost always aware of their artificiality. They are coming to us from screens, or speakers, or devices held in our hand.

But VR engulfs us. When we use even the most basic VR, we slide occluding goggles over our eyes and cover our ears with headphones, overriding our two primary sense systems with simulated digital signals. In more advanced VR we engage the body in the virtual experiences and create physical feedback from interactions with virtual objects. When this is done right, our brain becomes confused enough to treat these signals as reality. Wherever you fall on the "Does media influence behavior?" argument, I can assure you, VR does. There is plenty of research over decades performed in my lab and in others around the world that has demonstrated these effects.

For these reasons, VR is the apotheosis of every media fear and fantasy we've ever had. I could fill a chapter with the different types of dystopian scenarios that people ask me about after they've seen what VR can do. Will people stop socializing with each other in the real world? Can VR be used for mind control? Can you torture people with it? Will there be government surveillance? Corporate surveillance? Will it make people more violent? What about pornography? (Incidentally, the answers are: No. Kind of. Kind of. Probably. Almost certainly. Kind of. What about it?)

There is a sense from many people that VR is sinister—that it represents the nail in the coffin of a natural, social-oriented mode of human life that has been gradually dying away. Why will anyone want to exist in the real world when they can live an immersive fantasy life in VR? I think this view seriously underestimates real life. I'm with Jaron Lanier, who likes to describe the most amazing moment in VR as the moment when you take the HMD off and are flooded with the full gamut of subtle sensory inputs that VR can't capture—fine gradations of light, smells, the sensation of air moving on your skin, the weight and torque of the headset in your hand—these are all sensations that are incredibly difficult, if not impossible, to effectively simulate in a virtual world. In a strange way, VR helps you to appreciate the real world more. Yes, many people will be using VR porn, but it will never come close to the real thing.

YOUR BRAIN ON VR

One of the questions I get most often, whether from parents, reporters, policymakers, or even casual visitors to the lab, is "How does VR change our brains?" With new studies coming out on a daily basis advancing our understanding of neuroscience and psychology, it's a natural question to ask, particularly since we now have at our disposal powerful tools that can peer into the physical behavior of the brain. Chief among them is functional magnetic resonance imaging—fMRI—which scientists use to measure just about every psychological process there is, ranging from learning to memory to persuasion to implicit bias. So why not use it to measure the effects of VR?

Unfortunately, this is easier said than done. As anyone who has ever squeezed into the horizontal tube of an fMRI knows, you must remain absolutely still for the machine to obtain accurate results. Move too much and the technician will need to restart the session, and that means an extra 20 minutes of claustrophobia and cacophony.

I have read a number of grant proposals seeking a new way to do fMRI recordings to allow patients or subjects to move their bodies freely during measurement. This technology is likely years away. Other brain measurement techniques, such as administering an electroencephalogram, or EEG, in which small metal discs with thin wires called electrodes are placed on the scalp, allow for slight amounts of movement, but there is a tradeoff—the more freely the person can move around, the noisier the measurement tends to be.

Now, consider what we know about VR. What makes VR special, different from watching a 3D television set, is movement. Good VR is about walking, grabbing, and pivoting your head to see behind you. VR was created to allow people to frolic. But sudden movements tend to be the norm, not the exception, as lots of VR experiences leverage the natural flinch and fear responses to cliffs, spiders, and looming objects. And these quick, jarring movements in a large space are the natural enemy of brain measurement.

Of course, there are some experiments where people try to use fMRI to measure brain activation patterns. But as we go through them, keep in mind the tradeoff that exists in all of them. Either the subject can't really move and it's likely a misnomer to call the experience "VR," or the subject is moving around, and the brain measurements likely are coarse. The popular press will shower coverage upon the few studies that do report on measuring brain activities in VR, but what the journalists are calling VR is typically a stereo movie that is played, or a videogame where a patient uses a hand controller. Neither of these leverage "presence," the secret sauce of VR.

For example, in 2016 I coauthored a paper in the journal *Science* with Stanford neuroscientist Anthony Wagner and his post-doctoral fellow Thackery Brown.[6] In this study we sought to learn what is going on in the brain when it uses previous experiences to create a mental representation of a problem it is trying to solve— specifically what the role of the hippocampus is in this process. For example, when you get in your car to drive to a location you've been to before, you "map out" the route you are going to take to use as a mental guide as you drive. (At least, before we all got addicted to Waze and Google Maps.) The problem with fMRIs is that the subject must remain very still within a magnetized metal tube— which severely limits the kinds of activities they can perform. Consequently, most studies have not been able to provide interactive and perceptually rich experiences. But virtual worlds can be explored with a controller. This raised a question that intrigued me more than the role of the hippocampus. Would the brain react to the exploration of this virtual space in a similar manner it did a real space?

We exposed humans to a virtual maze and asked them to navigate it to five different locations. It wasn't as immersive as VR.

Subjects saw the maze through a screen, and they moved through it with a hand controller. The next day, the participants were required to plan and ultimately navigate to the same learned virtual locations while inside an fMRI scanner. During the latter session, whole-brain, high-resolution fMRI tests were completed as these individuals planned their route, as well as during actual navigation. Analysis of the data revealed that the hippocampus was engaged during goal planning. The orbitofrontal cortex, which is known to interact with the hippocampus during memory-guided navigation, plays a key role in classifying what constitutes a "future goal," or end point, in navigation.

While this was important for psychologists—to understand the neural underpinnings of how humans draw on past memories during the planning of future actions—there is another way to interpret these data. The interactive portion of the simulation, even though it was not as immersive as VR, still provided the richness to form a memory. My colleagues had very specific predictions about which brain areas would contain information about where an individual is in virtual space, as well as where they are next planning to go. Because it is technically impossible to examine the neural underpinnings of such navigational planning while people are actually moving through space, the use of non-immersive VR provided the means of doing so while people were lying still inside the fMRI scanner. The predictions were on track, in that the events experienced in virtual reality showed hippocampal activity during recall similar to what would be predicted for real experiences. Basically, even the non-immersive VR created a pattern of activation that neuroscientists predicted for actual events.

So one strategy for studying brain activity in VR is to use non-immersive VR. Another is to study animals, where one can use surgically invasive techniques (obviously not allowed with humans) to measure brain activity. In a 2014 study by UCLA scientists, researchers placed a small harness around rats and put them on a treadmill that allowed them to track walking movements (similar to the omnidirectional treadmills sold for VR video-games). The rats were surrounded by a virtual room on large video screens in a dark lab room. The scientists measured the activity of hundreds of neurons in their hippocampi. They also had a carefully crafted control condition where the rats walked in a real version of the room. Brain activation patterns were quite different in the VR and the real conditions. In the virtual world, the rats' hippocampal neurons fired chaotically, as if the neurons had no idea where the rat was—even though the rats' navigation behaviors seemed normal both in the real and virtual worlds. In the UCLA press release, Mayank Mehta, an author of the study and director of the W. M. Keck Foundation Neurophysics Center, claimed that "The 'map' disappeared completely. Nobody expected this. The neuron activity was a random function of the rat's position in the virtual world."[7] In addition to the random firing pattern, there was also less brain activity in the VR condition than the real-world condition. Mehta goes on in the press release to make a fairly bold claim: "The neural pattern in virtual reality is substantially different from the activity pattern in the real world. We need to fully understand how virtual reality affects the brain."[8]

This study was not without its critics. While most scientists focused on the generalizations from a rat brain to a human brain, VR experts had a different critique about the VR itself. A harness and a treadmill must have been a brutal experience for a rat, with tracking likely highly inaccurate. Perhaps the most likely explanation for the effect was simulator sickness. I have used a number of VR treadmills in my time, and I suspect my neurons were pretty scrambled as well—it's really hard on the perceptual system.

Other studies provide more nuanced accounts of brain activity and virtual environments. For example, in 2013 a study published in the *Journal of Neuroscience* simultaneously examined the size and complexity—that is, the number of decision points—of a virtual maze. Eighteen adults participated in the virtual navigation task, and then later on the subjects were put in an fMRI and shown screenshots of the mazes. Hippocampal activity was measured during these brief presentations of screenshots from the various mazes. Activity within the *posterior* hippocampus increased with maze size but not complexity, whereas activity in the *anterior* hippocampus increased with maze complexity but not size. This is called a "double dissociation" and is an important tool for brain scientists. It provides preliminary evidence that the brain has different areas that account for experiencing size and complexity of an environment.[9]

It is critical to remember that just about everything we do causes activity and changes in the brain, whether running across a field or eating a pizza. Not every change in brain activity is traumatic. For me, the critical question from a societal standpoint concerns the effects of prolonged VR use. The answer is, right now we just don't know. But it's not something we should treat lightly. Consider this description of a television show, in the process of being pitched to HBO, that I was approached to help produce.

"UNTITLED VIRTUAL REALITY SHOW is a groundbreaking experimental documentary project that takes 8 strangers and immerses them in a 30-day experiment in social isolation and virtual reality immersion."

The casting call for the subjects in the experiment was fairly intense, to say the least:

"VR PARTICIPANTS: 8 diverse people, 18–35 (age is flexible) who aren't afraid of grueling test of mental/physical limits. Participants will be kept in solitary isolation for 30 days and fed a bare minimum diet. Their only communication with each other or the outside world is in a VR world."

And no, you are not misreading that. They are being deprived of food and social contact with live humans and will interact with one another via avatars in networked VR. The producers asked me to embed experimental procedures into the show, and to help measure brain function in the participants. I actually took a day to think about it, I am embarrassed to say. It did seem like an important study, and one that might prevent a possible dystopian future if we learned about substantial brain changes given this type of horrific treatment. In other words, five years from now this description will likely be the norm for a subgroup of the population who will intensely overuse this technology. But I politely declined, as it just sounded like an unpleasant way to spend my day—depriving people of actual social contact.

Nonetheless, this is the type of study I get asked about quite often—what is the effect of replacing social contact with virtual social contact over long periods of time? But in my mind, this is akin to the dilemma that doctors often face. Nobody wants to run a controlled experiment where they force half of the subjects to smoke two packs per day and half to not smoke, to show a causal link between smoking and cancer. Instead we wait for correlational evidence, because ethically, few are comfortable with this type of study. (For the record I am also not ethically comfortable with tampering with rat brains to answer these questions.)

THE DOWNSIDES

In my Virtual People class, which I have been teaching since 2003, I always include a few lectures about the downsides of VR. As I hope to demonstrate in this book, VR experiences can be incredible, and truly offer promising new ways to address societal problems like poor educational outcomes, prejudice and discrimination, and inaction about the climate change crisis. But if a virtual experience is powerful enough to alter our fundamental views about the planet or race relations, then it must also have possibilities for ill. I'm going to briefly discuss four of those—experiential modeling of violence, virtual escapism, distraction from the world around us, and overuse. The critical question that faces us as we enter an era that will feature pervasive use of this deeply immersive medium is, how do we walk the line and leverage what is amazing about VR, without falling prey to the bad parts?

BEHAVIORAL MODELING

The concept of behavioral modeling was developed by Stanford psychologist Albert Bandura in the early 1960s, an outgrowth of his groundbreaking work on social learning, today one of the most studied ideas in modern psychology. Social learning theory posits that, under certain conditions, people imitate the behavior of others. Behavioral modeling is one facet of social learning, showing that just watching others engage in a behavior can cause a viewer to imitate it. This was a controversial argument at the time, as by and large the field believed that learning occurred by doing—that an actual reward or punishment was necessary for learning. Bandura made the pivot: we aren't just rats who learn by getting cheese after running through a maze; instead we often learn from watching others. Vicarious learning matters. Our understanding of this phenomenon began with the famous Bobo-doll study, conducted in the early 1960s in a Palo Alto nursery school.[10]

Two dozen young children watched an adult actor behaving

aggressively toward a popular toy called a "Bobo doll"—an inflatable clown figure weighted down with sand so that when it is punched it stands back up. The adults attacked the Bobo doll in a scripted manner. They hit it with their hands, kicked it across the room, tossed it in the air, yelled at it, and even smashed it with a hammer. There were two control conditions, each with two dozen kids. One group saw the same adult play nicely, ignoring the Bobo doll and playing with other toys. The second control group of children did not see any adult actor at all. Afterward, the kids got a chance to play with Bobo themselves. Those who watched the adult behave aggressively in turn became far more aggressive, relative to the other groups, in their own play. They hit the doll more, yelled at it more, and even created novel ways to be violent, for example using both a gun and a hammer to pound the doll with two hands at once. Every year I show a video of this original study to my class on "The Effects of Mass Media," which I have been teaching for more than a decade. It's difficult to watch, and my students often gasp out loud as some of the children really attack the doll in a way that seems like more than just playing.

Of course, there have been hundreds of studies since this original experiment that develop the theory, and determine which conditions make it likely for people to imitate others. Of relevance to this book is whether people imitate other people they see in media. Bandura, in fact, was interested in this question too, and two years after the first study, he and his colleagues replicated this Bobo doll experiment using film. Instead of watching a live actor beat the doll, children instead watch a video. Children who watched violent behavior toward the Bobo doll on TV were twice as likely to be aggressive toward it in real life as the nonviolent control group.

The brain science supports the idea of behavior modeling. In 2007 a group of psychologists ran a study in which subjects entered an fMRI scanner and watched a movie.[11] The ostensible purpose of the movie was to teach the subjects how an experiment they'd agreed to participate in would work. The movie featured an actor doing a learning task, and receiving painful electrical shocks when he made a wrong answer. After the movie, the subjects began the actual experiment and did the same learning task, believing they were about to get shocked when they got an answer wrong. The purpose of the study was to compare brain activation when subjects watched the training video to the brain activation that occurs when a person is actually doing the task and about to receive his own electric shock. The great part about this design is that subjects never actually got a shock—they just needed to believe they would receive one while they did the learning trials. Results demonstrated that the fear response, characterized by activation of the amygdala, was high in both observing and in doing. The brain activation pattern was similar in the two situations. In the words of the authors, "This study suggests that indirectly attained fears may be as powerful as fears originating from direct experiences."

Surprisingly, when compared to the plethora of violent, first-person shooter games that dominate much of the traditional video gaming marketplace, as I write this book there have not yet been a lot of these types of games released in VR. Time, of course, will tell. And it is revealing that the first-person shooter, *Raw Data*, one of the first hit VR games with over $1 million in sales in its first month, eschewed the gratuitous blood and guts of traditional videogames, choosing instead to pit the gamer against robot enemies. Many game designers, it turns out, quickly realized that there is a big difference in how we feel about performing violent actions on a screen versus performing them in VR. This is even truer when motion tracking technology is involved. Making an avatar commit violent acts by pressing buttons on a controller in a traditional videogame is an entirely different experience from when the same action involves pointing a gun with your hand at a three-dimensional representation of a person and pulling the trigger, or using your hands to strike or stab a virtual opponent in a violent game. Many designers who have experimented with making first-person shooters for VR have acknowledged that such visceral, gory gameplay is perhaps too intense for a large consumer market. "We made some core decisions early on that we weren't going to kill people," game developer Piers Jackson said about a first-person shooter he was working on. "Not having to confront death—that was something we deliberately chose. . . . You will feel it, like everything in VR, you will feel everything much more intensely."[12]

Perhaps in response to the absence of these types of games, it wasn't long after the first high-end consumer VR systems were released that hard-core gamers began modifying their favorite games to work within VR. One of the first examples of this that I saw involved the latest title in the infamous *Grand Theft Auto* series, famous for its dark comic sensibility and nihilistic violence. The game had been modified—"modded"—to allow users to experience the game world in VR. This included the ability to use VR motion controllers to point and aim the weapons in the game with the natural motions of the hands, rather than by pressing buttons on a controller or mouse. A short video of the player's brief adventures was posted online, showing a first-person perspective as the player punched a few computer-generated characters (who cartoonishly flew high into the air), shot a few virtual cops, and then drove down the street shooting and running over terrified pedestrians. The content was no different from that of the game as it might normally be played, but knowing that each of those punches, each of those gunshots, had been acted out by the player was disturbing. And in fact, in a blog post, the creator of the mod later expressed misgivings about his creation: "I'm really not sure about this," he wrote, under a GIF showing gameplay footage in which a taxi driver is shot in the head and the virtual passenger is shot in the back while he runs away. "I feel horrible about making this. You actually feel guilty. My mouth dropped the first time I shot someone in my *GTA: V* VR setup."[13] (It should be noted that in spite of the developer's ambivalence, he continued to develop his GTA mod.)

I have my doubts about how widespread the appeal will be for hyperviolent entertainment in VR. There will certainly be a subset of gamers who will enjoy immersive first-person violence, but I don't think it will be as large a group as currently exists with traditional games. I became even more confident of this prediction after experiencing my first taste of VR violence in a demo for the HTC Vive. Called "Surgeon Simulator," it is an incredibly well-designed demo of the Vive system's capabilities, with great graphics and interactivity. In the experience you perform an autopsy on an extraterrestrial in a space station above Earth, using a full tray of medical instruments, power tools, and weapons. The alien flails his arms as you operate, simulating flinch responses—even though he is ostensibly dead. Your tools begin to float around the operating room, governed by an incredibly detailed simulation of the physics in a weightless environment. So any object—whether it is a bone saw or a power drill or a Chinese throwing star—can hit any other object with the proper force, collision, and effect. We've probably demoed a few hundred people in this simulation, and as far as I can tell, the responses fall into two classes.

Some people, once they realize that the power drill actually can be driven into the alien's eyeball to produce a blood splatter, decide to not actually torture the alien. It's just not their idea of fun. Like the subjects in Slater's VR obedience study, they have a hard time inflicting pain on even an obviously fictional entity.

Others, however, don't seem bothered in the moment. They "have at it," slicing the alien up and exploring the dark, sadistic comedic potential of this gory scenario. (I even saw one use the alien's severed hand to hit him in the face.) The first time I chopped up the alien—I was transformed by how amazing the fidelity of the VR was—it affected me for hours. And not just being embarrassed at behaving in such a ridiculous manner in front of my students, but more emotionally. I simply felt bad. Responsible. I had used my hands to do violence. The experience of performing surgery on a lifelike entity stayed with me—I actually felt remorse hours after the experience.

What effects immersion in VR scenarios of fantasy violence will have on users is going to be a subject of serious discussion and study in the coming years. The subject is already highly politicized. Though a handful of loudly visible scholars have disagreed with the science, there is a substantial amount of psychological research showing that playing violent videogames increases arousal and aggressive behavior in players, and may increase levels of antisocial behavior.[14] Further research has shown that not only does playing violent games increase anger, but that violent gameplay on a 3D television increased the level of anger significantly.[15] Brad Bushman, one of the coauthors of the latter study, has been studying the effect of violent videogames for decades. According to him, "3-D gaming increases anger because the players felt more immersed in the violence when they played violent games. . . . As the technology in videogames improves, it has the ability to have stronger effects on players."[16]

The immersion created by 3D television is small compared to an experience in VR. When considering the demonstrated emotional and behavioral effects VR has on users, exposure to violent media should be a concern. Violent games are considered protected speech, as the Supreme Court decided in *Brown v. Entertainment Merchants Association* in 2011, a case which struck down a California law banning the sale of violent games to children without parental supervision. The reasoning of the judges was that the potential negative psychological effects of violent games were not strong enough to justify the limiting of free speech. But Justice Samuel Alito, in a separate concurring opinion, did distinguish between traditional media and virtual worlds, warning, "the Court is far too quick to dismiss the possibility that the experience of playing video games (and the effects on minors of playing violent video games) may be very different from anything that we have seen before. Any assessment of the experience of playing video games must take into account certain characteristics of the video games that are now on the market and those that are likely to be available in the near future."[17]

But there is another worry when we consider VR and behavior modeling, and that takes us back to the topic of Chapter 1: training. Independent of the hotly debated arguments about emotional effects and desensitization to violence, there is little doubt that VR can effectively teach the skills required to *succeed* at violence. During his trial, the mass murderer Anders Behring Breivik described hours of practice using a holographic sight in the popular first-person shooter *Call of Duty: Modern Warfare* as "training" for his horrific rampage on the island of Utøya on July 22, 2011.[18] Indeed, academic research has shown that playing violent games with a pistol-shaped controller, as Breivik appears to have done, does increase shooting skills like firing accuracy, and can influence players to "aim for the head."[19] For the reasons already mentioned, such training can only be more effective in VR.

None of this should come as a shock. These effects are why the military has flight simulators and uses VR to train soldiers for combat. It works.

ESCAPISM

With the notable exception of *Star Trek*'s holodeck, fictional accounts of virtual reality are routinely dystopic. In *The Matrix*, the movie's virtual world has been constructed by machines to keep humanity from realizing they are an enslaved species. In the popular novel *Ready: Player One,* the only refuge from a wrecked world suffering from gross income inequality and environmental destruction is a massive virtual universe into which people retreat whenever they have a chance. In William Gibson's *Neuromancer*, a novel I use as a textbook in my Virtual People class at Stanford, VR is essentially a medium for crime and prostitution. Its main character, Case, is so attached to the virtual world that he refers contemptuously to his own body as "meat," a fleshy prison that stands between him and "bodiless exultation of cyberspace."[20]

Virtual reality is depicted in these and other fictions as a place of ultimate escape, with disturbing consequences for the physical world. Based on contemporary behaviors, retreat into a self-curated fantasy world no longer seems like science fiction. How much time do we already spend with one part of our brain tethered to our mobile devices or computer screens, not only through social networks like Twitter, Snapchat, and Facebook (and whatever comes next), but also through videogames, Internet forums, and all other manner of entertainments?

Now imagine a world in which social media resembles the best party you were ever at in college, gambling online makes you feel like you're in the most exclusive room in Las Vegas, and pornography approaches the feeling of real sex. How many will be able to resist these instant gratifications for the inconveniences of "real life"?

Societal fears about escapism into nonreal worlds preexist the digital era, of course. Long before television was derided as the opiate of the masses, there was hysteria about novel-reading, and some students of media like to cite Plato's discomfort with the popularity of poetry in ancient Athens to dismiss the most recent bout of technopanic that surrounds the latest media invention. But I worry that VR will be of a different order of magnitude, that its difference from other media—particularly in terms of psychological impact—is one of kind and not degree. There is good reason to believe that in the not-too-distant future, much of what we do on the Internet will be done through virtual or augmented reality, further cutting us off from the physical world. The concerns about a "new solitude" that have been raised by thinkers such as Sherry Turkle will only intensify when the pull of these virtual spaces becomes ever more compelling.[21]

If VR indeed becomes a medium through which people access and interact with the Internet, it will only further increase our anxiety about how the Internet is changing millennia of human social norms.

OVERUSE

How long can people stay in VR at one sitting? In my lab, we have a "20-minute rule," which is not to use VR for longer than 20 minutes without a short break. It's fine to go back in, but take a minute, rub your eyes, look around, touch a wall. Spending a lot of time in a virtual environment can overwhelm the senses, and in my experience it's a good idea to rest for a couple of reasons.

Simulator Sickness
Increased computer processing power and improved HMDs have helped reduce some causes of simulator sickness, but using VR can still cause fatigue and discomfort. One of the main culprits is lag.

Imagine moving your head to look around a room, and seeing the world freeze in place or move in slow motion. Like a cell phone conversation that is constantly marred by delay, or a video that has the sound and image out of synch, this mismatch between what our perceptual system expects and what it sees is maddening for our brains, and for many people it has physical effects.

A sure cause of simulator sickness is a low frame rate. The first-generation HTC Vive runs at an impressively quick 90 hertz, but if a world is too detailed, or a computer is too old to properly render the scene, it can only redraw the scene for about half of those frames at 45 frames per second. The result is a jumpy scene that is hard on the perceptual system. It used to be worse. At old update rates of 30 frames per second, each frame lasts about 33 milliseconds, which means for that entire duration—an eternity for the visual system—the world is "frozen" and cannot update as it jumps from one position to the next. The problem becomes worse when you factor in another one-tenth of a second for latency, the delay caused for the physical processes of the tracking and computing systems to do their work. It may seem that computers work instantaneously, but our perceptual systems are fast enough to sense the time it takes for slower systems to register a physical movement, compute the newly updated scene, and then draw the pixels on the display. This is a second way to create lag, as the virtual world's movements correlate with physical movements, but not in perfect synchrony. Think of a digital clock that only updates once per second. The number of times it would update per minute is a good metaphor for the frame rate. But the time can also be behind—regardless of the update rate, each sample, in this case each second, could be ten minutes behind, consistently. That would be latency. Nowadays we update at 90 frames per second and the latency is so quick it's negligible, so this doesn't much contribute to discomfort.

For faster contemporary systems, latency occurs for a number of reasons. Most often, it is because a producer designs a VR environment to be so rich and detailed that it forces the computationally taxed computer to reduce the frame rate. In my lab we have a rule—we always run at the top frame rate. My programmers get very depressed when they have to simplify a scene, for example, to have 6 avatars in a crowd instead of 12 or to have the spheres be a bit blocky. But it's *always* the right strategy to reduce visual detail to ensure the highest frame rate. I see companies making this mistake often, and as someone who studies the brain and perceptual system it is heartbreaking to witness.

Eyestrain

The second reason I advocate for short times in VR is eye fatigue. Here is a test. Put your finger close to your eye, and then look at something far away. Even though your finger is in your field of view, it is blurry. Next, without moving your head, focus your eyes on your finger. Whatever you were looking at far away is now blurry, even though it is still in your field of view, while your finger is now clear and sharp. In the real world, our eyes do two things as we focus on objects. First, they converge, meaning that as a pair, our eyes move together to focus on objects, pointing inward for objects up close and more straightforward for objects far away. Second, each eye accommodates individually. In other words, each eye can focus similarly to a camera lens, and does so by changing the shape of the eye itself. In VR, it is fairly simple to converge—your eyes simply move to follow objects as they move from left to right inside the display. But accommodation works differently—the focus of each lens in VR is basically set by the headset maker. You can change the shape of your eye to your heart's content but you can't shift the focus inside of the image. Experientially, in VR, it doesn't matter where you focus—the sharpness of the scene stays the same due to the unchanging accommodation rates. There are only a handful of academic studies on this, and while the data are far from conclusive, most academics and thought leaders in VR believe this problem will prevent long-term use of headsets.

Reality Blurring

These physical limitations of VR haven't stopped some intrepid researchers and enthusiastic users from exposing themselves to virtual environments for long periods of time. More studies need to be conducted on the physical health consequences of such use, but there are already interesting anecdotal data about the how intensive use can contribute to a third potential hazard of long-term VR use, reality blurring.

Consider the German psychologist from the University of Hamburg who spent 24 hours in a VR room under carefully monitored conditions, and then published his findings in an academic journal in 2014. In the results section of their paper, the authors tell the scientific VR community: "Several times during the experiment the participant was confused about being in the VE (virtual environment) or in the real world, and mixed certain artifacts and events between both worlds."[22] These academics showed an academic's restraint in their description of reality blurring, but a YouTube personality named Derek Westerman, who holds a world record for time spent in VR, spoke more directly about his mind-boggling 25 hours in the matrix. Due to the *Guinness Book*'s strict rules, Westerman was only allowed to use one application. He chose *Tilt Brush*, which places the user in a black void in which he can create three-dimensional objects with controllers. "There is definitely a difference between my life before and after spending days in Virtual Reality. I was marked by it. And now, in an exciting way, everything feels slightly superficial or unreal." Perhaps not as scientific as the German psychologists, but a similar sentiment.

DISTRACTION

In my classes, I tell a joke that Stanford students' legs don't work unless their thumbs are moving. For many students, walking without texting simply is not an option. Visitors to Palo Alto are often stunned to see students texting while biking. It is not a rare occurrence. But imagine when we no longer need hands to communicate, and instead of seeing misspelled words in 140-character bursts, we are immersed in real social scenes, virtual ones, while trying to accomplish daily tasks. It is not a pretty picture.

In spite of what proponents of multitasking say, attention is zero-sum.[23] We only have so much of it to go around. And VR demands one's total attention. (Augmented Reality and Mixed Reality are another matter, and carry their own hazards.) Wearing a headset, and believing you are in a virtual world, can be dangerous to users and the people around them. In the months after VR was initially released, incidents of users hitting walls and ceiling fans, tripping over coffee tables, and even accidentally punching other people began to be shared in the media. Some of these incidents even made their way to YouTube, a new subgenre of comedy.

VR AND KIDS

By mid-2016, over $6 billion in investment money had been earmarked for VR, and the explosion of content that is created will include many games, educational programs, and VR experiences that will be aimed at children. It is astonishing, then, to think of how little studied and understood are the effects of immersive VR on children. How children react to media is of particular concern because their prefrontal cortex, the area that is associated with emotion and behavior regulation, is not completely developed.

The kind of reality-blurring that was described by the VR endurance athletes in the previous section may be a particular danger for children using VR, even in short bursts. Young children are, for instance, notoriously susceptible to acquiring false memories when exposed to everything from verbal narratives to mental images to altered photographs. In a study we performed at our lab in 2008, we tested how well kids in preschool and kids in early elementary school could differentiate virtual experiences from real ones, both directly after exposure, and one week after exposure to the virtual experience. For example, after giving kids a VR experience of swimming with whales, many formed "false memories," believing they had actually physically been to SeaWorld to see an Orca (as opposed to just seeing it in VR).[24]

Seeing the fast-approaching deluge of powerful VR content on the way, we decided to look at what effect immersive VR might have on children's behavioral and cognitive responses, by comparing the effects of interaction with a virtual character on either a television screen or in immersive VR. With the help of the Sesame Workshop, the creators of *The Muppets* and *Sesame Street*, we programmed a simulation in which children from the ages of four to six could interact with a virtual character—in this case the loveable, furry blue monster Grover. Fifty-five children participated in the study.

When compared to the non-immersive VR condition (watching Grover on a television screen), children in VR showed a significant deficit in inhibitory control, as measured by their success playing a game of Simon Says with Grover. At the same time, children who experienced immersive VR were more likely to interact and share with the character in VR.[25]

This study suggests that immersive VR elicits from children different behavioral responses than we have seen from television, and that children may process the immersive VR content differently from other media. In the real world, young children often have a hard time controlling their impulses, especially in response to something tempting. That is what makes Simon Says such a fun game. The child simply wants to mimic the adult even if she doesn't say the magic phrase. On TV, the temptation is not so hard to resist. Children do quite well at NOT mimicking the gesture when Grover doesn't say "Simon Says." But in VR, the temptation is salient, just as in the real world. All the things that make VR special—that it is immersive, responds to natural body movements, and blocks out the physical world to make the illusion that much more compelling—make virtual temptations hard to resist.

Manufacturers of HMDs have been careful to acknowledge the potential dangers of VR for children, and many have age restriction guidelines and warnings about use. Sony's PlayStation VR, for instance, suggests that no one under the age of 12 should use their product—a big ask, considering the entertainment ecosystem in which it exists. PlayStation's use of playful, cartoony avatars and its large catalog of child-friendly characters and games make this warning seem fairly academic in nature.

As I like to tell journalists, uranium can be used to heat homes and it can also be used to destroy cities. In the end VR, like all technologies, is neither good nor bad—it is a tool. And while I am fascinated and bullish about the amazing experiences that will be shared, the social possibilities it will open up, and the creativity it is going to unleash, we cannot ignore the darker side. The best way to use it responsibly is to be educated about what it is capable of, and to know how to use it—as a developer or a user—responsibly.

This is because most of us, in some way, are spooked by the ever-increasing virtuality in our daily lives, as we spend more time in communicating online, consuming Internet media, and staring at our phones or tablet computers. Virtual reality represents the culmination of the way technology is enveloping our lives, and our uneasiness. These invariably depict VR to be such an irresistibly seductive medium, so good at giving users direct access to their personal dream worlds, that they lose their taste for the challenges and frictions of real life, choosing instead to retreat into a disconnected fantasy life.

The philosopher Matthew Crawford, best-selling author of *Shop Class as Soulcraft*, addresses this fear in his book *The World Beyond Your Head*. It is a thoughtful take on how digital technologies are encroaching on our relationship with the material world, and affecting our ability to select the things that are important and pay attention to them. "As our experience comes to be ever more mediated by representations, which removes us from whatever situation we inhabit directly, as embodied beings who do things, it is hard to say what the principle of selection is. I can take a virtual tour of the Forbidden City in Beijing, or of the deepest underwater caverns, nearly as easily as I glance across the room. Every foreign wonder, hidden place, and obscure subculture is immediately available to my idle curiosity; they are lumped together into a uniform distancelessness that revolves around me. But where am I?"[26]

An excellent point. It is important to be mindful of the limitations of a technology like VR, particularly in light of the extravagant claims often made about it. Crawford is concerned about the ability of digital simulations to smooth over the wrinkles in life, believing that escape into virtual spaces disorients us and weakens our ability to appreciate or function in the physical world. This can certainly be true. But is there a way that VR can enhance our connection with the real world and with other people? What if we used VR to deepen that appreciation by expanding our ability to empathize with others' viewpoints, or to get a better understanding of other experiences, or to understand the consequences of our behavior to other people and to the environment we all share?

CHAPTER 3

WALKING IN THE SHOES OF ANOTHER

We walked for days crossing the desert into Jordan. The week we left my kite got caught in a tree in my yard. I wonder if it is still there. I want it back. My name is Sidra, I am 12 years old. I am in the 5th grade. I am from Syria in the Daara Province, Inkhil city. I have lived here in the Za'atari camp for the last year and a half.[1]

These words mark the heartrending beginning of the virtual reality documentary *Clouds over Sidra*, an eight-and-a-half-minute-long 360 immersive film that takes viewers inside the Za'atari refugee camp in northern Jordan, home to over 80,000 Syrians displaced by civil war. In a voice-over you hear the young girl Sidra describing the camp, while different scenes of daily life in the sprawling complex play out around you. You see her with her family in the small converted shipping container in which they sleep. Next, you are in a makeshift gymnasium where men are lifting weights to pass the time. Then you are among a group of young men, laughing and talking as they bake flat bread in a large open oven. Then you are standing in the middle of a soccer field, while young girls kick a ball around you. All the while, through a translator, Sidra matter-of-factly narrates scenes of normalcy amidst this ongoing humanitarian tragedy.[2]

The first time I saw *Clouds over Sidra* was at the Tribeca Film Festival in April 2015. I was struck by the sense of intimacy the film created between viewers and the film's subjects. In one scene, for example, children line up for school in one of the muddy streets within the camp. Temporary dwellings stretch out beyond sight in all directions. As I looked around, I could see and feel a vastness that the limited field of view offered by a photograph or filmed images simply cannot convey. Eighty thousand is just another heart-numbing statistic—an abstraction—until you are standing in the middle of this enormous makeshift city in the middle of the desert. Some of the children, clearly amused during the filming by the large, strange-looking camera apparatus that has been plunked down in the middle of the street, approach it, laughing and making funny faces, or just scrutinizing it. It felt like they were interacting with me. Many of these children had walked to Jordan through the desert, fleeing death and destruction, just as Sidra had, and the reality of that fact, combined with their playful normalcy, is intensely emotional.

When the experience was over, I noticed others around me wiping tears from their eyes as they removed the Samsung Gear headsets from their faces. The film was clearly unlocking powerful emotions. But these reactions hadn't been produced by a dramatic score, or clever editing, or close-up shots that lingered on a particularly poignant face or detail. These aspects of traditional filmmaking, designed to intensify our emotional engagement, are virtually absent in this immersive film. Viewers of *Clouds over Sidra* are simply confronted with a series of ordinary moments—people baking bread, families laughing together, children at play and in school. The major difference is that the immersive video made us briefly feel as if we were there with them.

Cocreated by artist and filmmaker Chris Milk, working with the support of Samsung and the United Nations, *Clouds over Sidra* is not just an early experiment in VR documentary, it is also an example of explicit VR advocacy. "We're taking these films," Milk said in 2015, "and we're showing them at the United Nations to people that work there and people that are visiting there. And we're showing them to the people that can actually change the lives of the people inside of the films."[3] This was from a talk titled "The Ultimate Empathy Machine," and in it Milk expressed his conviction that the immersive properties of VR make it particularly suited to sharing the experiences of others, to deepening our understanding of lives outside our own. VR, Milk has said, "connects humans to other humans in a profound way that I've never seen before in any other form of media. And it can change people's perception of each other. And that's how I think virtual reality has the potential to actually change the world."[4] Indeed Milk's predictions came to fruition; according to the United Nations, the VR experience has literally doubled the number of people who donate.[5]

Milk's work in VR has reflected his commitment to this belief. One of the premiere directors in this new medium, Milk has gone on to create several VR short documentaries, including *The Displaced* for the *New York Times Magazine*, which describes the experience of three children from Syria, Ukraine, and Sudan as they attempt to rebuild their lives in the wake of war. The *New York Times* distributed over a million Google Cardboards to subscribers to accompany this featured piece of VR journalism. With these short films, Milk and some other early VR content creators join a venerable tradition in the history of art, using their medium to encourage empathic understanding. A social novel like *Uncle Tom's Cabin,* which employed narrative storytelling in text and illustrations to impress upon northern audiences the suffering of slaves in the years before the American Civil War, is just one of countless examples of political art. Others include Francisco Goya's *Disasters of War* series, which captured shocking images of violence and conflict in the early nineteenth century (although they weren't widely shared until decades later), or Jacob Riis's photographs of New York City slums published in his book *How the Other Half Lives*. For as long as humans have created art there has been a preoccupation with understanding and conveying the experience of others, particularly suffering—we see it in painting, sculpture, photography, film, and more recently even in certain videogames.

Whether all this empathic art has actually changed human behavior is a subject of ongoing debate. Some have pointed to statistics showing large declines in war and violence throughout the world and suggested that this is in part due to improvements in communication, which have broadened our innately tribal outlooks and contributed to an "expanding circle" of moral concern.[6] As a consequence, our instinct to favor our family or tribe expands to include previous outgroups. There is an impressive amount of data to broadly make this case, but also significant exceptions to this view that the power of our rational minds, informed by empathic experiences, is leading to undeniable progress for humanity. Some, for example, point to recent examples of societal fragmentation and terrible violence, as seen in Nazi Germany or in Rwanda in 1994. In both these cases, modern communications media were actually used for propaganda to demonize others.

Empathy has been the subject of much psychological research and debate in recent years, but many psychologists agree that two different systems, psychologically distinct, appear to be at work when we experience empathy. One is an emotional system which reacts reflexively when we encounter the suffering of others. It is

activated when you flinch upon viewing footage of an athlete getting injured, for instance, or when you turn away from a particularly gruesome scene in a horror film. The second major ingredient
of empathy is cognitive, the ability of your brain to form theories
about what other people are feeling and what might be causing
those feelings. For some psychologists, the basic model of empathy ends there, but I don't think that goes quite far enough. For
instance, psychopaths, con men, torturers—people in these categories might be very good at the emotional and cognitive aspects
of empathy, but still lack the prosocial quality we usually associ-
ate with empathic people. For this reason I favor the conception
of empathy put forward by my Stanford colleague Jamil Zaki, who
adds a third component to what he calls "full-fledged empathy": a
motivational element. Does a person, after experiencing the emotional and cognitive reactions to another's suffering, have a desire
to alleviate the suffering of that person? Does a person expose herself to a trying empathic emotional experience in the first place?[7]

Zaki's empathy model is also illuminating because it highlights
how empathy is something we can *choose* to experience, rather
than something that just happens unconsciously. In Zaki's view,
we turn empathy on and off because feeling it can be emotionally
taxing—a drain on our mental resources. As an example, Zaki asks
us to imagine a scenario in which we are watching television and we
learn a leukemia telethon is about to come on, featuring children
suffering from the disease telling their story. Few would doubt that
this experience would create empathy in viewers. But there would
be competing motives over whether to watch it. In some situations,
you may be curious to learn more and continue watching. In others, you might want to avoid feelings of guilt, or just the sadness
that hearing the stories might cause you.

Another example: most of us don't spend the mental energy

required to imagine what it is like to sleep out on the street every time we pass by a homeless person. If we did so, it would be difficult to function, given how traumatizing that understanding would be for most of us. Instead, we usually ignore the homeless person, or push the depressing reality of his existence further out of mind by constructing a story about the person that makes it easier to dismiss their suffering, perhaps by attributing their plight to choices that individual might have made. It is sad that this person is homeless, we might tell ourselves, but he shouldn't have used drugs or gotten fired from his job. We put aside the many reasons beyond that person's control that might have put him on the street—a financially crippling medical condition or mental illness, for example.

Choosing to experience empathy is not without costs. In intense situations it can cause real lasting pain and psychological damage. We see this in the damaging psychological effects suffered by certain professionals, such as doctors, nurses, psychotherapists, and emergency workers—people who confront extreme human suffering on a regular basis. Psychologists call this "compassion fatigue," the strain caused by pushing our empathy to its limits. It can lead to symptoms like anxiety, nightmares, dissociation, anger issues, and burnout.[8]

Still other occupations may require the suspension of empathy in certain aspects of the job, especially ones that are extremely competitive. Politicians, soldiers, and professional athletes all have good reasons to switch empathy on and off. And none of us are immune—we all live in a world with stunning inequality and many demands on our emotions. It is the rare person who can make it through the day without ever consciously putting on blinders to the suffering and inequality around us.

Empathy isn't a fixed quality. Our capacity for empathy can be changed by our culture and the media technologies that transmit the culture's values—for good and bad. Certain types of training can desensitize people to other people's feelings, even as other types can enhance our empathy. Research has shown that one of the best ways to foster empathy is the psychological process of "perspective-taking," or imagining the world from another person's point of view. Perhaps the scholar most associated with this work is Mark Davis, a psychology professor who has done most of his teaching and research at Eckhard College in St. Petersburg, Florida. Davis created the most popular tool for measuring perspective-taking, a questionnaire called the "Interpersonal Reactivity Index," which asks people to rate their agreement with statements such as "Before criticizing somebody, I try to imagine how I would feel if I were in their place," or "When I watch a good movie, I can very easily put myself in the place of a leading character."[9] In 1996, Davis also published what is likely the canonical paper showing how perspective-taking actually works. The title of the paper tells it all: "Effect of perspective taking on the cognitive representation of persons: A merging of self and other."[10] His work over the past few decades has shown that if one imagines the world from another's perspective, the gap between oneself and the other decreases. Thinking similarly to another person literally causes changes in cognitive structures, such that one's thoughts concerning the other become more "selflike."

Consider a famous study from the year 2000 by Adam Galinsky, now a professor at Columbia Business School, showing that perspective-taking causes empathy. Experimental participants— college students—looked at a photograph of an elderly man and were asked to write an essay describing a day in his life. One group of the participants were given no explicit instructions; they were simply told to write about the man's day. The second group were

given specific "perspective-taking" instructions and were specifically urged to take the first-person perspective of the individual
in the photograph when writing their narrative essay. Compared to
the control condition, perspective-takers showed more empathy.
They took longer to recognize prejudice phrases associated with
negative stereotypes in a psychological reaction-time task, and
also expressed more positive attitudes toward the elderly when
writing their essays. But how many people will actually be able to
use this imagination technique in their daily lives?[11]

Taking into account VR's ability to create seemingly real experiences and to allow users to look at a virtual world from many
different points of view, one would imagine that VR would be particularly suited for perspective-taking. First, it relieves users of the
cognitive effort required to make a mental model of another person's perspective from scratch. This might help a user overcome a
motivational hurdle that causes her to avoid taking perspectives
of others.

This leads to a second advantage of virtual reality in enhancing empathy. Because the mental model of the perspective of the
empathic subject can be created in great detail in VR, it can be
designed to help avoid stereotypes and false or comforting narratives. For example, if a teenager has a negative stereotype of the
elderly, then merely asking the teenager to imagine an elderly
perspective might simply serve to reinforce those stereotypes.
The teen might create images of elderly who are slow, frugal, and
who tell boring stories. But formalizing the role-playing into a set
simulation that counteracts stereotypical representations can
avoid these negative stereotypes by creating scenes that show the
strengths of the elderly. People might not have the right information to actually take a proper perspective, and VR can guide them
through the process more accurately.

No medium, of course, can fully capture the subjective experience of another person, but by richly evoking a real-seeming, first-person perspective, virtual reality does seem to promise to offer new, empathy-enhancing qualities. We can read news accounts or watch a documentary about refugees, but these media require a lot of imaginative work on our part. Narrative can give us a lot of information about life in the camp, but it is poor at conveying what living in a camp feels like. We don't have a mental library of the appropriate sights, sounds, and story to imagine what it is like to be a refugee. VR can convey the feeling of the camp's environment, the smallness of the living quarters, the size of the camp. It can bring Sidra, and the other members of the camp, alive in a way that documentary footage can't.

It seems that the world has taken as a given that VR can foster empathy in unprecedented ways. But does it actually work? What does the research tell us?

THE VIRTUAL MIRROR

My lab at Stanford has been running and publishing experiments on VR and empathy since 2003. We have looked at ageism, racism, helping the handicapped, and other domains. And what we have discovered is that the answer to the question of whether VR is, indeed, "the ultimate empathy machine" is nuanced. It's true that VR, across studies, tends to outperform control conditions. But it is not a magic bullet. It doesn't work every time, and the "effect size," the statistical term for how strong an outcome is, varies. The first study we ever ran, published in 2006 but designed and conducted starting in 2003, looked at ageism, the bias that young people tend to have toward the elderly. We were inspired by the Galinsky study in which people looked at a photograph of an elderly man, and imagined what it was like to see the world from his eyes. But we decided to take imagination out of the equation.

First, we designed a "virtual mirror" that allowed a subject to see her avatar reflected on the wall of the virtual version of the lab room. We asked each subject to walk up to this virtual mirror, spend about 90 seconds gesturing in front of the mirror, and closely observe her "reflection" as it moved with her. She would roll her head left and right, touching her ears to her shoulders. In the virtual mirror, her reflection would do the same. Then she would take a step forward physically and watch her mirror image get bigger. After some time, we would ask the subject to crouch outside of the mirror's frame so she could no longer see herself, and then to pop back up and watch her reflection rise with her.

VR mirrors had been around since long before I came on the scene. In fact, a virtual mirror is one of the demos that sold me on VR in the late 1990s. I was visiting UCSB for the first time and experienced the scenario in Jack Loomis's lab. Jack is one of the pioneers of VR, and he had long been using VR to study distance judgment, vision, and other aspects of perceptual psychology. But Jack had an intellectual curiosity that extended far beyond the human visual system, and his VR demo suite reflected that. In 1999, the mirror was fairly low-tech—a user's reflection was a clunky, robotic avatar, and only rotated its head and moved back and forth without having any hand or leg movements. But it still created a powerful effect. With enough time in VR, that clunky body began to feel like your body. It was a glimpse into the future, and one of the first experiences I built when I arrived at Stanford was the mirror.

The principle at work was based on the "rubber hand illusion." This now classic experiment was first performed in the 1990s by two scientists at Princeton University who had their subjects sit at a table, positioned so that one of the subject's hands was hidden out of sight, underneath the table. Instead of seeing his actual hand on top of the table the subject would instead see a lifelike rubber one where his actual hand should be. Neuroscientists then gently stroked both the hidden hand and the visible rubber hand simultaneously with a paintbrush. If the two hands were stroked synchronously and in the same direction, then the subject began to experience the rubber hand as his own.[12]

How was this determined? One example was that when asked to point to his hand, most of the time a subject would point to the rubber hand instead of the one hidden under the table. Interestingly, if the strokes are not timed well, then the illusion fails. But if the touch is synchronous, then the hand gets incorporated into one's body schema. For example, if the neuroscientist subsequently threatens the rubber hand by making a stabbing movement toward it with a needle, fMRI scans shows activity in the part of the subject's brain that is normally activated when a person anticipates pain, along with the area that is normally activated when a person feels the urge to move his or her arm. This landmark study using rubber hands has informed us how to induce body transfer into avatars. For example, if someone sees his avatar get lightly poked with a stick, and also physically feels his chest getting poked synchronously, the avatar is treated as the self. People "transfer" their consciousness into it, according to dozens of studies. This mirror technique is now a fairly commonplace one—neuroscientists today call it "body transfer."[13]

When one sees her avatar—whether it's from the first person looking down at her digital body, or by looking in the virtual mirror—and events occur synchronously, her mind takes ownership of the virtual body. The brain has very little experience—from an evolutionary standpoint—at seeing a perfect mirror image that is not in fact a reflection of reality. Because we can digitally create any mirror image we want in VR, VR offers a unique, surreal way to allow people to take ownership of bodies that are not their own. Because the brain incorporates the mirror image into the self, the self is malleable. The mirror is the ultimate tool to allow someone to become someone else, to walk a mile in his shoes. One's mirror image can look just like her, or be bigger, become male, grow a third arm, or even change species.

In our study on ageism, we made the mirror image older.[14] First, we induced body transfer in a group of college students by having each of them perform a series of motions as he looked at his avatar in a virtual mirror. The avatar matched his movements, as you would expect a mirror image to do. However, half the college students were looking at an avatar that resembled them in age and gender. The other half saw an avatar of the same gender that was clearly in his 60s or 70s. Once we had induced ownership of their new body, the students then turned away from the mirror and met a confederate—a person networked into VR with them, who was part of the experiment. From the subject's point of view, there was simply another avatar—this one of college age—in the room with him, and the two avatars could talk and move about the room together. We very clearly reminded subjects that the image they saw in the virtual mirror was exactly how they would appear to others in the virtual environment. In other words, from the subject's point of view, the confederates would believe the subject *was* older and had no idea the avatars didn't reflect their actual, young age.

The confederate then asked the subject to step forward and answer some questions to spur some social interaction, saying things like, "Tell me a little about yourself" or "What makes you happy in life?" In order to highlight the social roles in the simulation, it is critical to have a personal interaction. Finally, the subject was asked to perform a brief memory exercise of trying to repeat a list of fifteen words. The purpose of the memory exercise was to reinforce the fact that he was wearing an elderly avatar, as one of the more prevalent stereotypes of the elderly is bad memory. We found it particularly powerful when a college student has his memory questioned by another young avatar while wearing an elderly body. The idea is to induce mild reminders of the negative stereotypes so that the subject gets to "walk a mile" in the shoes of another.

In the study, the subjects who wore the elderly avatar used more positive words to describe the elderly in general, compared to those who wore a younger avatar. For example, they were more likely to list the word *wise* than *wrinkled* when pressed to list the first thing that came to their minds. We looked at three common measures of bias, but only found significant expressions of ageism in one of the three measures, the Word Association Task. This is a previously developed psychological technique during which subjects were asked the open-ended question, "When you think of somebody old, what are the first five words that come to mind?" Then, two coders who were not a part of the study rated the response as having a positive or negative connotation. The difference was about 20%. This was a modest but noticeable improvement. While our results did not show a consistent effect across all three dependent measures, it was still encouraging to find that such a short virtual interaction—less than 20 minutes—can change a person's negative stereotypes at all.[15]

A few years later, in 2009, working with a doctoral student named Victoria Groom, we used the virtual mirror to examine race.[16] Victoria thought that racial empathy would be induced when white participants wore black avatars. She reasoned that if a white person took the perspective of a black person, her negative racial stereotypes would break down. Groom ran about one hundred participants, half of whom approached a virtual mirror in a black avatar and half in a white avatar. This time, the confederate performed a job interview, where the subject had to answer questions about aptitude and previous work experience.

As in the ageism study, we looked at three separate, previously validated measures of bias. And again, just as in the ageism study, only one of them—in this case the Implicit Association Task, which gauges reaction time to positive and negative concepts—showed a significant difference. However, we noticed that inhabiting a black avatar had the opposite effect from what we observed in the ageism study—it actually caused people to score *higher* on standard measures of implicit racial bias than those who wore a white avatar. In other words, wearing a black avatar primed more racial stereotypes instead of creating empathy. Stunningly, this pattern was true not only for white participants but for black participants as well.[17] Regarding virtual racism, it seems that the story is complicated, and that wearing a black avatar actually reinforced stereotypes and made them more salient.

In the fall of 2016, I had a conversation with renowned Harvard psychologist Mahzarin Banaji, who helped invent the Implicit Association Task and is likely the world's expert on implicit racial bias. She had read this study and wanted to discuss possible future interactions. Her feedback on the study was clear—she would have predicted the same outcome, given our procedure.

Social psychological research has demonstrated that indicators of a stereotype's social group such as physical features indicating gender or race can activate concepts relating to those social groups. Often these concepts are widely held stereotypes—in many cases negative—that can directly affect cognition, attitudes, and behavior. For example, the well-documented stereotype held by some Americans that blacks are prone to violence has been demonstrated in a number of studies. This implicit bias happens automatically, without intention or awareness. One explanation for our findings is that we didn't succeed in encouraging perspective-taking. Given the limitations of the early technology, we were using a very simple hardware system that did not track arm movement and only allowed rudimentary body transfer. It is hard to be certain if body transfer had actually occurred. Perhaps we didn't induce transfer but instead simply primed the negative stereotypes that subjects already had about race, which is why we produced the counterproductive results.

But it's not all bad news.

Mel Slater and his colleagues in Barcelona ran a similar study a few years later. Slater has some of the best VR technology in the world, and is literally the world's expert at inducing body transfer into avatars. In his system, he tracked the body movements much more accurately and fully than we did in our early study, and consequently demonstrated with a high probability that body transfer was induced. In his study, being embodied in a black avatar *reduced* bias among white participants as measured by the Implicit Association Task, compared to control conditions. Slater and colleagues concluded that "embodiment may change negative interpersonal attitudes and thus represent a powerful tool for exploring such fundamental psychological and societal phenomena."[18]

Perhaps the strongest case of this powerful effect can be seen in the work of Sun Joo Ahn, a former graduate student of mine who now is a professor at the University of Georgia. In a paper published in 2013, Sun Joo conducted three experiments to explore whether becoming colorblind in VR would improve empathy toward those who are colorblind.[19] Subjects first received basic information about red-green colorblindness. They then entered VR to do a sorting task designed to be particularly difficult if one could not differentiate perceptually between red and green. Half of the participants wore a headset with a colorblind filter applied over the objects on the screen, which allowed them to accurately experience being red-green colorblind. The other half also wore the HMD and did the sorting task but only imagined that they were colorblind.

In one of the experiments, after leaving VR, subjects who became colorblind spent almost twice as much time helping persons with colorblindness compared to participants who had only imagined being colorblind. The helping task involved participants working with a student group that was supposedly trying to build colorblind-friendly Web sites. The task required them to view screenshots of Web sites and to write about why the Web site might be inaccessible to colorblind individuals and how it might be improved. It was made clear to the participants that this activity was not a part of the experiment and that it would be volunteer work. We measured the time spent volunteering and showed that VR increased helping.[20]

Informal reactions from participants were telling. "I felt like a colorblind person. I felt like I was in a whole different world. It made me realize how tough it is for them to do certain things in life, such as drive." This points to a unique strength VR has in sharing with users some of the challenges people with physical disabilities face. It is a powerful experience to inhabit the avatar body of a minority and experience a scenario in which you are discriminated against, but a VR scenario can't hope to capture all the subtle aspects of discrimination a person experiences in her life. But it is easier to illustrate the difficulties a person with perceptual or physical disabilities might have, although again, it's important

not to overstate findings that have not yet been extensively replicated. Research has shown that we still must take care with how
these experiences of disability are implemented. Experiments by
Arielle Michal Silverman into empathy for the blind, for instance,
have shown that the initial disorientation caused by making a seeing subject blind might actually promote discrimination instead
of empathy, because the newly blind subject experiences only the
trauma of sudden blindness, rather than the realities of being
blind for a long time, and all of the acquired skills that go along
with it. Instead of perceiving the blind as empowered, able people,
a person in this study might only focus on the difficulties of coping
with the sudden loss of sight.[21]

We also sent the colorblind participants a questionnaire
designed to gauge their opinions toward colorblind people 24 hours
after the study. We demonstrated that, for people who in general
have a hard time feeling concern for others—that is, people who
scored low on the Interpersonal Reactivity Index described above—
becoming colorblind caused more favorable attitudes toward colorblind people than imagining it. This is preliminary evidence
that VR is a great tool for those who tend to have a hard time being
empathetic. For people who were "naturals," that is, those who
in general tend to engage in empathy, there was no difference in
attitudes 24 hours later. But for those who struggle with empathy,
the virtual tool allowed them to improve their perspective-taking
abilities.[22]

EMPATHY AT SCALE

In 2014 a program officer from the Robert Woods Johnson Foundation was on a lab tour and experienced some of our empathy
demonstrations. I explained to her the general methodology of

the research—inducing body transfer followed by an experience. She was intrigued, but skeptical. We began a conversation about the robustness of the effects. As should be clear from reading this chapter, most of the time VR empathy seems to work inside the lab. But how strong are the effects? How long do they last before the treatment wears off?

The conversation evolved into a three-year project called "Empathy at Scale," where we sought to test the results of our studies and attempted to understand how well this tool works out in the real world. In the first series of experiments, published in 2016, we sought to discover the boundary conditions of VR empathy. Finding boundary conditions is a popular strategy in validating experimental findings; basically psychologists change the conditions as they reproduce experiments to discover when their effect stops replicating, looking for the "ceiling." Very few findings generalize to every single situation in life, and understanding when a tool no longer works is important information. In particular, we wanted to look at threat. Perspective-taking and empathy are most important, arguably, in situations where there is tension. I get a few calls per month from folks suggesting to use VR to bridge the gap between warring nations. In these situations, there is a preexisting and deep tension; psychologists call this "threat."

We returned to ageism and once again used the virtual mirror to embody our subjects as elderly. In the first study, subjects either imagined being older, similar to the Galinksy study described earlier, or became older in the mirror. In addition, we also varied threat, high or low. In the high threat condition, before entering VR, subjects read a passage called "Elderly Pose Immediate Threat to Young Americans." In the low threat condition, they read an article titled "America Prepared for Changing Demographics." Both were about the implications of the demographic trends of people living longer, but one framed them as a threat, while the other framed them as a situation America was prepared to deal with. We then made the manipulations more impactful by asking subjects to write an essay to reflect upon the demographic trends and how they would affect their own lives. To sum up, there were four conditions. Half the subjects imagined being older, while half became older. Within each of those conditions, half were threatened by the elderly while half were not.

In this study VR succeeded in buffering threat. For those who imagined being elderly, being threatened caused them to have less empathy for the elderly. This is what we would predict; it is difficult to demonstrate empathy for those who might harm us. But for those who became elderly via body transfer in VR, the opposite occurred. They actually had more empathy when they were threatened compared to those who were not threatened. One possible explanation for these results is that it may have been easier for participants to take the perspective of an elderly person when they were in VR compared to when they were relying entirely on their imagination. Similar to the colorblind study, VR may be particularly beneficial when the intergroup context makes it difficult for people to engage in perspective-taking.

But now, enter the boundary effect. We ran a second study where we made the threat personal. Instead of having the subjects read a passage about the elderly, we created two elderly avatars and had the subjects interact with them in VR. We used a classic ostracism task developed by psychologist Kip Williams called Cyberball. In this exercise, three people are to play catch with a ball, and as the play progresses, one of them is left out—nobody throws him the ball.[23] It may sound trivial, but there have been dozens of experiments using this task and the feeling of being ignored is significant. Ostracism hurts—some studies have shown in an fMRI that

Cyberball ostracism causes the areas associated with pain to light up. Others simply show it makes people feel saddened.

In our study, we again had the same four conditions—imagination versus VR, and high and low threat. Regardless of condition, after the perspective-taking exercise, subjects then put on the headset and played a ball toss game with two partners, both of whom were clearly elderly. In the high threat condition, subjects received a total of 3 tosses out of 30 tosses. In the nonthreat condition, they received a total of 10 tosses, which is what one would expect in a game with equal sharing.[24]

Results demonstrated that the threat manipulation was about twice as strong in this study as in the first study, when subjects simply read about the elderly. Those who only received three throws felt angry and offended according to self-report measures. Also, in almost all of the measures of empathy, subjects who were threatened were later less empathetic to the elderly in general compared to those who were included in the ball-tossing game. But the critical data concern the effect of VR. In contrast to the promising results in Study 1, in the second study that had more intense, experiential, and intentional threat, there was no difference between VR and imagination. In other words, increasing immersion was not enough to overcome empathic avoidance. Rather, when under threat, participants consistently showed negative attitudes toward the elderly. Comparing the two studies, this preliminary evidence suggests that perspective-taking via VR can be effective at fostering positive behaviors toward the outgroup when the intergroup threat is indirect, but not as effective when the threat becomes more concrete and experiential.

All of the studies I've described in this section, like the vast majority of all psychological studies, have a major limitation. The elephant in the room is that we psychologists present findings taken from a limited and narrow sample group and make conclusions as if they apply to all people. However, the statistics we perform dictate that we can only draw inferences specifically about the sample used in our observations. As a result, a majority of studies in the field of psychology only truly inform us about upperclass, college-educated people below the age of 22 who are taking an "Introduction to Psychology" course.

In truth, research findings that appear to be robust can crumble when taken to the general population, as the foundational work of Stanford psychologist Hazel Markus—who has made a spectacular career out of showing cultural differences in psychological processing—has demonstrated. For example, during graduate school, I worked with both psychologists and anthropologists to take a "bedrock" psychological finding out into the real world. We used a reasoning task that is taught in cognitive psychology textbooks across the world, specifically, how people rely on similarity-based notions of typicality and central tendency in order to perform natural categorization. Consider the following two arguments:

Robins have sesamoid bones, therefore all birds do.
Penguins have sesamoid bones, therefore all birds do.

Which argument is stronger? Undergraduates will tell you the first one, and most readers will likely agree. Robins are the more typical bird, so a trait from them is more likely to generalize to all birds. But in our research, the preference for typical arguments failed to replicate when looking at those with high domain knowledge—for example, birders from the Chicago area and tribal members of the Itza Maya in Mexico, who interact with birds more often than college undergraduates.

There is reason to suspect that empathy in particular will vary according to demographic. Jamil Zaki recently developed a frame-work to understand motivation for empathy. He focuses in par-ticular on individual differences. At least three motives—trying to avoid emotional suffering, incurring material costs such as hav-ing to donate money to help, and worrying about competition by performing weakly at work or in social situations—drive people to avoid empathy. Similarly, at least three motives—doing a good deed, which results in positive affect; strengthening affiliation with in-group members such as friends and family; and wanting others to see one as a good person, called "social desirability"—drive them to approach empathy. To truly understand how a VR empathy treatment works, it is important to have a large enough sample size to seek out populations that would likely vary in these traits.[25]

Consequently we have developed the "Empathy at Scale" proj-ect, with the goal of running 1,000 subjects through our study to see how well VR works at promoting empathy, compared to typi-cal media techniques like using narrative and statistics. The 1,000 subjects are unique not just for the magnitude of the sample, but also in the diversity of the sample. We have installed VR systems on the road—our Mobile VR Unit—in museums, near libraries, and at festivals and fairs to try to get people who are not just the typical college students.

As of September 2017, across two studies, we have data from over 2,000 subjects. It's been an adventure, and we actually published an entire paper summarizing the lessons we have learned from VR field studies, ranging from how to accommodate those who can't stand up for long periods of time, to which direction the line to the tent should form, to how to alert people in a public place that we are running a study. One of the more interesting differences between college students and nonstudents is the amount of time it takes to complete a questionnaire. Students race through them, while non-students spend more time reading and answering. After pilot testing the study in the lab with college students, we thought it would only take 25 minutes. But in the field with "real people," it takes about 45 minutes.

Our simulation is called "Becoming Homeless," and premiered as a VR film at the Tribeca Film Festival in April 2017. Users experience the transition from having a job and a home to losing them and becoming homeless. In the experience, you endure the cascading series of events that often lead people to become homeless. First you lose your job and are forced to sell your belongings. Unable to pay rent, you are evicted from your apartment and are forced to sleep in your car while you continue to look for work. Then, one night, sleeping in your car, you are roused by police and fined for violating a local ordinance. The car is sold, and you next find yourself harassed by a stranger while trying to get sleep on a bus. We designed the experience not only to be moving and compelling, but also interactive. For instance, you are forced to choose between which items you want to sell in your apartment—the sofa, the television, your phone? We ask the user to navigate the confining space of the small car he is living in to brush his teeth. On the bus, you must watch your bag and fend off strangers who approach you as you try to sleep.

With 900 subjects run, we have begun to look at the results. In general, VR is working better than the control conditions, which involve reading a narrative about becoming homeless or reading statistics about the homeless. When compared to these controls, people who have had the VR experience have shown more empathy on questionnaires, and are more likely to sign a petition to support affordable housing. But as in the other studies, the effect is not uniform across dependent variables. The effects are consistent across about half of our measures but are by no means definitive. It seems that VR is the best medium of the four we chose, but the effect size is modest. Of course, we are continuing to analyze the data.

VR AND EMPATHY APPLICATIONS

In 2003, I arrived at Stanford as an assistant professor. One of the biggest priorities was to get my new lab funded, and being in Silicon Valley, I turned to industry in addition to the traditional routes of the National Science Foundation and other government sources. The first grant I got from Stanford was a small gift fund from Cisco to study social interaction in VR.

The executive from Cisco was a big thinker, a woman named Marcia Sitosky. Her tour of the lab caused a light bulb to go off after she saw herself in the virtual mirror and viewed our empathy research. She urged me to think about using VR for training in diversity. So much effort is spent on diversity training in corporations and other organizations, but for all the resources, the methods fall short. Could a VR experience be the answer?

In 2003, I was head down, tunnel-vision focused on publishing academic work, which meant I had to forgo many great opportunities to build applications. Between 2003 and 2010 (when I received tenure), I published dozens of experiments that used the virtual mirror. Most of them were centered around the Proteus Effect, namely, that when one wears an avatar, he implicitly *becomes* like that avatar. People in taller avatars negotiate more aggressively, people in attractive avatars speak more socially, and people in older avatars care more about the distant future. This theoretical work sought to understand the psychological mechanisms of how avatars changed the people that wore them.

But in the back of my mind, Marcia's suggestion resonated. While in the lab, she reflected on how corporate diversity training was far from perfect. While the general idea was promising—having people learn about workplace harassment with the goal of preventing it—as implemented it was not very effective.

Indeed, the research backs up Marcia's intuition. A 2013 paper published by Harvard sociologist Frank Dobbins examined the existing data on diversity training. Dobbins and his coauthor concluded that "initiatives designed to quash managerial bias, through diversity training, diversity performance evaluations, and bureaucratic rules, have been broadly ineffective."[26] My own experience resonates with this assessment. At Stanford, every eighteen months I am asked to participate in one of two training methods. I either watch a presentation by an acting troupe in a large group, or I do an online assessment—driving-school style—where I read about case studies and then get tested on the legality of various behaviors. Each is better than nothing, but in my opinion, neither is changing the way I think, other than perhaps giving me some good advice on how to pragmatically deal with harassment in the office.

In 2003, the last thing I wanted to do was build corporate training software. I needed to work 80-hour weeks to publish papers, and distractions like this were not allowed. But my life post-tenure was really about scaling our lab's work. Stanford is a university that encourages us to face outward and deliver our research out into the world.

Adam Silver, the National Basketball Association commissioner, visited the lab in 2015. He came, with the rest of the executives from the NBA, during a technology tour of Silicon Valley. His initial motivation was to investigate using VR to allow fans to watch games from their living room, as if they were courtside, or even better, as one of the players. I politely let him know why I didn't think this was a great idea—I think two hours in an HMD is a bad idea. And we'll further discuss the trials and tribulations of how to show narrative and information via VR later. But what really grabbed the group—especially Eric Hutcherson, the head of human resources for the NBA—was diversity training. We spent a solid hour in a conference room discussing possibilities, and the group was extremely excited to move forward.

A few months later, Roger Goodell, the National Football League commissioner, came to the lab expressly to talk about diversity training. The NFL, like the NBA, brought most of its executive committee. The massive corporation known as the NFL was surveying Silicon Valley technology to find ways to improve the fan experience with their product and improve the league. One of the best reactions from that visit came from their chief information officer, Michelle McKenna-Doyle, who got to experience becoming an NFL quarterback by doing a dozen plays in our QB training simulator. When she came out, she had an interesting reaction. She argued that she had long felt that, conceptually, she knew as much if not more about the game than any of her male colleagues, but they could dismiss her insights because "she has never been on the field before." With VR, she felt that she'd had that experience, and she left the lab truly feeling she had a new understanding of the game.

Former NFL quarterback Trent Edwards, who is also a cofounder of STRIVR, has actually made an interesting point related to McKenna-Doyle's. He notes that allowing fans to see the QB's perspective might actually improve their understanding of what is happening on the field, and perhaps give them a better appreciation of how difficult the job is. Athletes are people too, and even though they draw huge salaries, the volume of hate mail and threats they receive could be reduced by perspective-taking.

Over the past few years we have built and tested a diversity training system to implement in the NFL. The biggest supporter within the organization has been Troy Vincent, a former NFL cornerback and current executive vice president of Football Operations. When we visited Troy in 2016 at the NFL headquarters to plan this project, one of the things I was most impressed with was his vision for how to tackle the problem. Instead of starting with the players, Troy suggested we start at the top of the organization, with league executives, owners, and head coaches. The way to change a corporate culture—and the NFL is a huge organization—is at the top, so we decided to begin there. We have built and tested with NFL personnel an interview simulation that is designed to train skills to avoid racial and sexual bias. Trainees get to relive the simulation multiple times to practice how to manage the inherent bias we all have. The idea is to leverage "repetitions" to give executives the tools to manage if not overcome their biases. The system debuted in February 2017, where scouts practiced interviewing draft prospects just before the NFL combine, and soon will have a larger role in the NFL. In an interview with *USA Today*, Vincent described his vision: "We'll start using this as another teaching tool later this year. We want to be known as the best place to work."

THE (FURTHER) WIDENING CIRCLE

The capacity of human empathy extends beyond other humans, of course. Many individuals prize animals as much as—if not more than—other people, and will lavish their pets with expensive food and health care. As we've seen the circle of empathy expand over the centuries, we've seen a growing number of species that were once casually abused or mistreated become deserving of moral concern, although for those who argue for animal rights there is still a very long way to go. Perhaps the most significant amount of animal abuse occurs in our system of factory farming, an industry driven by our strong demand for cheap meat. It was concern over this system that led to one of the most intriguing experiments in VR empathy we performed in our lab, led by one of my students in 2013.

Joshua Bostick was one of those rare undergraduates who insisted on calling me "Professor Bailenson" instead of Jeremy. A Master's student, he had worked in the lab on and off for a few years as an undergraduate, and after learning the ropes, began to help out on research projects and in ushering public tours of the lab. During this time, Joshua transformed from a research assistant into a scholar. As I got to know him I noticed he was inspired, perhaps even fixated, on a single idea: putting humans into the avatar of cattle in order to reduce meat consumption and the environmental consequences caused by our high dietary demand for beef.

When he first made his pitch, I was taken aback by the boldness of the idea, and politely suggested he go back to the drawing board. Though I make an effort to cut down on my meat consumption, I still enjoy a burger from time to time, and I didn't want to look like a hypocrite. "So do I," he replied, and he told me he loved steak. Eventually he convinced me that a study that embodied humans in cow avatars would not be about trying to convert people to vegans, but instead would seek to reduce the amount of meat people eat—resulting in less energy use, deforestation, and carbon dioxide production associated with raising cattle. He reminded me of the fascinating work of Temple Grandin, who introduced humane reforms into cattle slaughterhouses that reduced the stress and panic of animals before they were killed, changes that she credited to her ability, perhaps enhanced by her autistic mind, to imagine the experience from the animals' perspective. Joshua's passion and persistence paid off, and after a few meetings, he convinced me to green-light the experiment.

For the next few months we pushed the envelope on the lab's technology—one of the challenges of doing VR research is that we have to do a lot of programming and engineering just to set up the studies. We created a system that could track, in real time, the arm, leg, back, and head movements of people as they crawled in our lab on all fours. Participants would wear a head-mounted display that allowed them to see themselves in the third person (or third cow?) as they walked about. In the meantime, every millimeter that their arms and legs moved was tracked and transformed into their cow's gait. In order to pull off this feat we designed a special vest a person wore that had LED tracking lights attached to it, and also bought knee pads to prevent subjects from chafing on the lab carpet.

To "become" the cow, people saw the cow they were controlling through the head-mounted display and heard farm sounds from the speaker array that surrounded them. We also made it so they felt virtual touch. When the subject saw his cow avatar getting poked by a cattle prod, he felt that poke in his actual side accompanied by a shocking sound from the direction of the prod. The illusion was completed by a vibration from the ground, created by "buttkickers"—low frequency speakers hidden in the floors—that rounded out the simulation of getting shocked. The cattle prod was not just for effect—as we saw with the rubber hand illusion, the sensation of touch enhances body transfer. If a person *feels* his chest being poked, and *sees* a stick poking his avatar, this facilitates the mental transfer from self to avatar.

The experience we ended up with was persuasive and stunning. We put 50 college students into the cow simulator. They experienced a day in the life of a cow, drank from a virtual trough, ate virtual hay, and finally were prodded to a truck headed for the virtual slaughterhouse. Compared to others in a control condition, who just watched the cows walk around and get shocked, those who became a cow gained more empathy for the plight of cattle using the most common questionnaire to measure self-reported empathy. But more telling than the formal analyses were the spontaneous quotes provided by the subjects. Here are a few of them: "Horrifying to be a cow and be poked by the prong," one participant wrote, "I was constantly on edge with what I would be forced to do next and what would happen when I ran out of things to do." Another remarked that it "made the horrible and sad lives of livestock animals seem more real and less theoretical than when I read about it." The body transfer technique clearly accomplished its goal, as pointed out by one student, who said, "I really did feel like I was the cow and I really didn't like being prodded." Even with the older technology—the study was run long before the much-improved consumer-grade headsets had arrived—the subjects experienced high presence: "It was surprisingly realistic. I felt as if I was in the place of the cow." It was clearly an intense experience.[27]

Our goal with this research was to discover ways for people to be better connected with where our meat comes from. Farming and slaughtering animals now happens far away from us, the meat we eat tidily packaged in ways designed not to remind us that what we are consuming was once a living, breathing animal. With our empathic imaginations stunted by this distance, it is no surprise that we overconsume and waste so much of the meat we produce. Joshua's experiment helped subjects imagine the animal's suffering and sacrifice.

While we were, in fact, advocating that people should consider eating meat a bit less often, we weren't trying to turn the United States into a Vegan Army, as some members of the conservative press suggested. Love the idea or hate it, becoming a virtual cow struck a chord with the public. Journalists—for example, the BBC, the Associated Press, Fox News, Yahoo, *The Daily Mail*, and newspapers from around the world—continue to call the lab to cover the study, and for every cattle rancher who has protested the idea, five others have applauded it, including restaurant owners, government officials, educators, and parents.

We can see how VR can foster empathy by enhancing our experiences of traditional perspective-taking techniques, but Joshua Bostick's cow study also shows us how VR's ability to introduce surreal experiences, such as allowing us to inhabit the embodied perspective of another species, can influence attitudes. As the use of VR grows, and we understand better how to leverage its unique affordances, we can expect to see more unusual and novel applications that open our minds and hearts—not only to other people but to ourselves. This was shown in an unusual study done by Caroline Falconer at University College of London and Mel Slater at Barcelona's Institution for Research and Advanced Studies.

People who suffer from depression are often intensely self-critical, unable to give themselves the patience and understanding that they might grant to other people. Falconer and Slater set up a scenario to see if VR could increase "self-compassion," and their research has shown some promising results.[28]

The scenario they studied worked like this. A depressive patient was placed in VR, where they interacted compassionately with a virtual child. The words the patient spoke to the virtual child were recorded. Some of the patients then reentered the scenario, but this time as the child, hearing their own comforting words spoken back to them as if from another person. Others heard their words from a nonembodied third-person perspective. While self-criticism decreased in both conditions based on scales designed to measure self-compassion, self-criticism, and fear of compassion, the study showed a "significantly greater increase in self-compassion" in the embodied condition.[29]

Typically, to foster self-compassion, therapists might encourage individuals to perform imaginative exercises, for instance, to think of how they might treat a friend who is going through a challenging experience, to write letters from the perspective of a friend to themselves, or to role-play as a criticizer and a person being criticized. In Falconer and Slater's elegant experiment, these more abstract exercises are turned into an immediate encounter that appears to amplify the traditional therapeutic technique. We will see this principle in effect in later chapters, which discuss the use of VR in a therapeutic setting. For now, I want to stick with the theme of perspective, and how changing one's point of view can have profound effects on her mental relationship to the world around her.

CHAPTER 4

WORLDVIEW

Edgar Mitchell was enjoying a rare moment of uninterrupted reflection as he gazed out of the Apollo 14 command module on its way back to Earth. He was nearing the end of the nine-day mission to the moon, and it had been packed with constant activity—scientific experiments, equipment monitoring, piloting the lunar module, collecting rocks, not to mention fraught hours dealing with equipment malfunctions during the descent to the surface. Once there, Mitchell, along with Apollo 11 veteran Alan Shepard, had also gone on two moonwalks, including one when Shepard famously hit golf balls with a club he'd fashioned from a 6-iron head and a lunar excavation tool. Mitchell's primary duties on Apollo 14 were to perform science experiments during the trip to the moon and on the surface, and he was also the pilot of the landing module. This meant that for the return trip, most of his work had been completed, giving him some precious time to relax and stare out a window.

The pilot of the command module, Stuart Roosa, had put the spacecraft in BBQ mode—a slow spin that evenly distributes the heat of the sun across the surface of the ship. As a result, through the window Mitchell was given a rare view. "Every two minutes," he later recalled, "a picture of the earth, the moon, the sun, in a 360 degree panorama of the heavens appeared in the spacecraft window." Mitchell, who passed away in early 2016, was one of only a handful of people who have been privileged with this extraordinary view. As Mitchell watched our planet, moon, and sun float past the window over and over again, he was struck by a powerful feeling of connection with the universe. He meditated on this experience for hours. Later, he would try to describe what he felt in an interview, saying, "I'd studied astronomy, and I'd studied cosmology, and fully understood that the molecules in my body, and the molecules in my partners' bodies and the molecules in the spacecraft, had been prototyped in some ancient generation of stars. In other words, it was pretty obvious in those descriptions, that we're stardust."[1]

After he retired from NASA and the US Navy a few years later, he founded the Institute of Noetic Sciences, an organization devoted to studying human consciousness. He would cite his experience on Apollo 14, and the unique perspective it gave him, as the inspiration for the Institute. "You develop an instant global consciousness," he later said of the view from space, "a people orientation, an intense dissatisfaction with the state of the world, and a compulsion to do something about it. From out there on the moon, international politics look so petty. You want to grab a politician by the scruff of the neck and drag him a quarter of a million miles out and say, 'look at that you son of a bitch.'"

Mitchell was not the first, or the last, astronaut to report a consciousness-altering experience after seeing the Earth from space. So many astronauts and pilots have been affected by this rare perspective that there is a name for the phenomenon—the "Overview Effect." Astronauts like Chris Hadfield, Ron Garen, and Nicole Stott have all reported versions of the sensation, a sudden intense awareness of the fragility of the Earth that sinks in when one sees how thin the Earth's atmosphere is, or first realizes that signs of human despoliation, like deforestation and erosion, are easily visible from orbit. Others have discussed being struck by the lack of boundaries when looking at Earth from orbit, the sense that the Earth is an organic whole and not carved up into a collection of nations, an idea that we internalize from seeing the Earth in maps or globes. On the international space station it is common, according to Hadfield, for astronauts to spend much of their limited free time staring out of its cupola window, an activity known as "Earthgazing."[2]

The right viewpoint can upend the way we see the world. Maps, bird's-eye-view paintings, diagrams of human anatomy, or the Copernican model of the solar system—each of these has profoundly changed the way we see ourselves and our place in the world. More recently, well-known photographs of Earth from space, like the "Pale Blue Marble" picture taken during the last of the Apollo missions, have been credited with raising public awareness of the environmental fragility of the planet. But those who have actually seen Earth from space know that images alone, no matter how powerful or iconic, are poor substitutes next to the experience. A frame cannot capture the infinite vastness of space, and the shock of seeing the Earth as insignificant like a tiny grain of sand in a vast ocean. There is a special power in being present. After all, there are plenty of pictures looking out from the rim of the Grand Canyon or visiting Victoria Falls in Zimbabwe, but that doesn't keep millions of people each year from going through the inconveniences required to see these sights for themselves. Anyone who has had the privilege of seeing one of the planet's natural wonders in person understands this.

If certain extreme perspectives can have this kind of psychological impact, and if VR can more powerfully convey the experience of those perspectives, how might VR be used to raise awareness about the planet? I began thinking along these lines in 2009, around the time I was being considered for a tenured position. For much of my first two decades of research, I had focused almost exclusively on social interaction. From a professional perspective, this made sense. I had started in psychology and then moved to the field of communication, so it seemed prudent to study social interaction, especially because the subject is well understood by government granting agencies and academic journals, and there is a wealth of practical applications for research in this area. Much of this work is described in my previous book, *Infinite Reality*, and in the intervening years these studies have influenced the fields of politics, marketing, education, and health care. Even during the years and years I spent testing aspects of social behavior like collaboration, persuasion, learning, and personality in VR social settings, I always knew that I wanted to expand the scope of my research to see how people interacted with virtual simulations of nature. This was largely because I believed those interactions might be useful in educating people about the consequences of what I believe to be the greatest peril facing our planet: climate change.

Like many of us, my concern about the damage humanity is doing to the environment has grown as my awareness of the science of climate change has increased. By 2010, I was downright alarmed. Contrary to impressions you might have about a scientist who spends his days tinkering with computer-generated environments, I have a long-standing love of the outdoors. I grew up in upstate New York where wandering around the woods with friends, looking for frogs and salamanders and snakes, was my version of Facebook.

But it wasn't just that. Stanford University is one of the leading institutions in the world studying climate change, and as a professor there I have been able to spend time with some of the all-time environmental greats—people like Steve Schneider and Chris Field, major figures in the development of climate change science and policy, and Paul Ehrlich, whose controversial books on the environmental costs of human population growth were so alarming they shifted public debate and action. In the first decade of the twenty-first century, as public awareness began to grow about the mounting evidence of a warming planet, I stood on the sidelines as climate change was debated and documented by my colleagues who wrote reports for the Intergovernmental Panel on Climate Change. Like many, I watched, stunned, as the recommendations of environmental scientists were repeatedly ignored by decision makers who either denied that climate change was caused by humans, or believed it wasn't happening at all.

By 2010, this disconnect was literally keeping me awake at night. In spite of the near-consensus of experts, and mountains of information, a large portion of the public remained unwilling to acknowledge the problem, let alone consider the daunting changes in policy and behavior that would be necessary to mitigate the damage. I thought VR could help. However, the lab up until this point had tunnel vision on social avatars, and it would be difficult to make that shift away from an area that we specialized in, especially one that provided reliable funding. Fortunately, around this time I got tenure, and with it the freedom to pursue research that might be less popular, or more difficult to fund, but which might have immense positive applications. It allowed me to shift the lab's vision—the grants we applied for, the experiments we ran, and the papers we wrote—to investigate what role VR might play in reversing the damage we are doing to the environment.

One of the first experiments we did came out of a 2009 *New York*

Times article by Leslie Kaufman, which described how the comfort and convenience of high-end toilet paper has huge environmental costs, not only for the world's old-growth forests and the animals that live there, but also for the Earth's atmosphere.[3] In her article, Kaufman reported that the use of soft, fluffy toilet tissue had exploded in popularity in the United States, and that creating this type of tissue was leading to the harvesting of "millions of trees . . . in North America and in Latin American countries, including some percentage of trees from rare old-growth forests in Canada." The loss of these trees reduced the absorption of the greenhouse gas carbon dioxide, a major factor in climate change. Moreover, some estimates indicate that over 10% of the pulp used in the United States comes from virgin forests, which, in addition to being irreplaceable in their own right, often provide critical refuge for endangered species. And the loss of trees is not the only cost: processing soft toilet paper uses more water, often involves the use of polluting chlorine bleach, and produces more tons of waste that takes up more space in sewage pipes and landfills.

There is an easy solution to this problem: using toilet paper made out of recycled materials. At the time Kaufman wrote her article, however, only 2% of Americans did so, compared to much higher rates in Europe. And yet despite this well-reported and convincing article, little in the intervening years has changed. It remains a sad fact that it is incredibly difficult to change our consumption behaviors, particularly in a world where the environmental costs of basic creature comforts like soft toilet paper are obscured from our experience.

But how can people be encouraged to see the ramifications of such a small, seemingly inconsequential decision as buying nonrecycled toilet paper? Psychological research into the critical factors that influence environmental behaviors has shown that some people possess a strong sense that their behaviors have a direct impact on the external world—what is called an environmental "locus of control."[4] These individuals are more likely to behave in environmentally conscious ways. Could we strengthen an individual's sense of connection to the environment in our lab?

Sun Joo Ahn, a graduate student who had spent years studying "embodied experiences," wanted to see if powerful effects displayed in our experiments on learning and persuasion would also apply to the environment. For her dissertation project she turned Stanford students into loggers. First, she constructed a virtual forest, one that was filled with trees, singing birds, and foliage. By all accounts, it was one of the most pleasant and convincing virtual worlds we had ever constructed in our lab—so much so that people who entered the virtual forest reported feeling calm—as if they'd stepped into an inviting glade on a comfortable spring day.

But Sun Joo was interested in more than a virtual vacation. She incorporated into the experiment a virtual chainsaw, which was actually a motorized haptic device duct-taped to a constructed saw-handle that would be handed to the subject just before the HMD was placed on the subject's head. Once the subject entered the virtual forest, she was asked to look around. When she looked down, she saw her virtual arms clothed in a logger's jacket and work gloves, and in her virtual hands she held a chainsaw. Subjects were then told to approach a nearby tree. Once they were facing it, the chainsaw roared to life, vibrating in their hands, and they were instructed to move it back and forth to cut down a tree in front of them. The virtual chainsaw slowly cut through the tree trunk and after about two minutes, it sliced through. The room then echoed with a loud crackling noise as the tree began to fall. The sequence culminated with the tree crashing to the ground. After the tree fell, the subjects were free to look around the virtual forest, now silent as the birds had all fled. The cut tree was splayed on the ground to their left.

It was a powerful simulation of an event that exists far outside of the lived experience of most people, reinforced by audio, visual, and even physical touch feedback. If information about the environmental consequences of our paper consumption were tied to such an experience, we wondered, would it have a stronger effect than if the consequences were described in print? In this first study, 50 experimental subjects were given statistics Sun Joo had computed based on the amount of toilet paper an American consumes, about 24 rolls per year. They were told that using nonrecycled toilet paper for the duration of their lives would cause the felling of two full-sized, standing trees. Then, half of the subjects read a carefully crafted written account describing what it would be like to cut down a tree, while the other half experienced the virtual tree-cutting. Subjects then filled out a questionnaire, were thanked, and left, believing that the study was over.

But Sun Joo concocted a way to encounter these subjects 30 minutes later, so that she could examine their behavior outside of the lab—in particular their paper use. Visibly pregnant at the time, Sun Joo used her condition, and the sense of courtesy it inspired, to assist in her experimental subterfuge. As the former subjects walked by her, she would knock over a glass of water with a well-rehearsed sweep of her arm. She then asked the subjects to help her clean up the spill, motioning toward her belly, and a stack of napkins, as she did so.

All of the subjects helped clean up the spill (apparently chivalry is not dead), but those who participated in the virtual tree-cutting used 20% fewer napkins than subjects who only read a description of tree-cutting. Having the virtual experience caused a noticeable increase in actual conservation behavior in a task thought to be unrelated to the study, providing evidence that an experience is more powerful than other methods of communicating environmental consequences.[5]

But how long does it last? Sun Joo designed a follow-up study to examine this question. She also used a more compelling control condition—instead of reading about cutting down a tree, half of the subjects watched a video, shot from the first-person perspective, of the activity. A week later, she polled people on their recycling habits, and found that after a week, subjects in the VR condition reported they were more likely to recycle than those in the video condition. In other words, not only did VR produce larger changes than other media, those changes were more enduring.[6]

A LUMP OF COAL IN YOUR STOCKING

Cutting down virtual trees is an analog to a real experience— logging an actual tree. But virtual reality also allows people to embody "impossible" or surreal experiences. Have you ever consumed coal? Not just used it to heat your house, but actually *eaten* it? Crunched it with your teeth, gulped it down, and coughed out coal dust?

My colleagues at Stanford received a grant from the US Department of Energy and were charged with designing virtual simulations to reduce energy use. They approached me to brainstorm how we could use my lab to create an experience that was compelling enough to change habits, and we spent a few weeks polling Stanford students, looking at their energy use patterns. We decided to focus on showers. If you have ever lived with a teenager, or were one once, you might remember that young adults really enjoy taking long showers.

But heating and transporting water is expensive from an energy standpoint. We computed, in consultation with environmental engineers at Stanford, how much coal would be needed to heat a shower. A ten-minute-long shower, we determined, used just under four pounds of coal. Assuming one showers daily, then by cutting down on the amount of energy—for example, by turning off the water while you lather, or simply by spending less time in the shower—the savings can really add up as the years go by.

We wanted to alert people to how much energy they were consuming in the shower. This is not a new approach to energy conservation: shower meters have been available for purchase for some time. For example, Shower'n'Save is one type of electronic meter that hangs in the shower and tells people how much money they are spending on heating water. For meters like this, the details about the energy you are using are displayed as numbers on a screen: they are informative, but perhaps not quite vivid or compelling enough to change behavior. We wondered if the ability of VR to convey this kind of information in a bizarre or dreamlike way would make a greater impression upon people.

In our experiment, subjects took a virtual shower. They put on the head-mounted display and were transported into a tiled bathroom with a showerhead and handles. As the virtual water poured down, steam rose and the floor vibrated. Subjects rubbed their physical arms, torso, and legs as if they were taking an actual shower, and inside VR they received alerts as to how much energy their shower was taking. We varied the nature of these alerts. Sometimes they were numbers on a virtual display (low vivid condition), and at other times the information was conveyed in a more intense way— by showing consumed coal stacking up in the virtual shower (high vivid condition). In addition, we varied how personal the feedback was. In the high vivid, high personal condition, the consumption of coal was depicted literally, with the subject depicted as eating it. In the low personal condition they saw the used coal stacked in front of them. In the text condition (low vivid), bathers read a virtual display modeled after a typical shower sensor that read "*You have used four pieces of coal*" (which was the high personal condition) or "Four pieces of coal have been used" (the low). In vivid conditions, subjects could see through a window just outside of the shower. Outside the window, either the coal stacked up onto a plate as the shower proceeded, or the bather saw a three-dimensional avatar designed to look like him reach down onto the stack of coal and pop a piece into his mouth. The avatar would then chew on the coal and cough. The physical floor shook the subject's feet as his avatar coughed in the simulation.

After the study, the subjects were required to wash their hands using a special sink that measured the amount and temperature of the water used. None of them thought it was part of the experiment—they all believed the cover story that everyone who uses the VR equipment must wash his hands. But subjects in the vivid conditions—the ones in which they saw the virtual coal—used less hot water than ones who only received textual cues. Just as in the tree-cutting scenario, receiving a visceral reminder in the form of coal was more effective than simply receiving words on a screen.[7]

These studies provided some valuable insights into how virtual reality experiences could teach us how our individual behavior was connected to environmental damage. They also showed that virtual reality could be more effective in changing behavior than other media like text or video. But there was still a lot more work to do. My initial goal was to improve education and awareness of climate change, and that was a much thornier problem. Climate science is extremely complex, and its worst effects won't begin to be felt for decades. There's a tremendous amount of misinformation out there, some of it being spread consciously. Then there's the simple fact that the factors contributing to climate change seem so entrenched, and the changes required to reverse it so difficult to enact, that many people prefer to pretend it's not happening rather than to do something. Like a person who has sunk deep into debt, it's psychologically easier to ignore the problem, or engage in magical thinking, than to start making the hard changes required to turn things around.

Frankly, even those of us who rail against the ignorance (or worse) of those who deny the potentially catastrophic effects of climate change could be doing a lot more.

In 2013, I attended a keynote speech by the former head of the National Oceanic and Atmospheric Administration (NOAA), Jane Lubchenco, at an event hosted by the Stanford Woods Institute for the Environment. In the address, Lubchenco, then in a visiting post at Stanford, discussed some of the lessons she had learned running the NOAA, the US government scientific body that oversees various agencies responsible for monitoring and studying the conditions of our oceans and atmosphere. Among its many missions, NOAA is tasked with protecting "life and property from natural hazards."[8] Her tenure at the agency coincided with the most extreme four years of weather yet recorded, so it was not a surprise that her talk was a moving—and at times dispiriting—keynote, containing heartbreaking details about all the natural disasters which she had encountered and managed responses to on the job, including six major floods, 770 tornadoes, three tsunamis, and 70 Atlantic hurricanes, not to mention record snowfalls and serious droughts.

One of the reasons I was drawn to Lubchenco's address was that it wrestled with the very subject I was dealing with in my lab: how hard it remains, over 25 years since the Intergovernmental Panel on Climate Change released its first report, to convince the public, and more importantly policy makers in Washington, that global warming is contributing to some of the extreme weather she had witnessed.[9] Then Lubchenco said something that really got my attention: she noted that as she visited communities that had been ravaged by natural disasters and interacted with the survivors, those who were suffering in the aftermath of these powerful weather events were more likely to believe the scientific consensus on climate change. "When people directly experience something," she said, "they see it in a different light."[10] When you feel the effects up close, when they become personal, they can no longer be ignored.

I had a chance to speak to Lubchenco a few years after the Stanford event. I told her what an impression her speech had made on me and my research. During our conversation, I was able to ask her about this question of the impact of experience: Could she describe one of these instances when a weather event changed someone's mind about climate change? She told me a story about a visit in 2011 to a city that had recently been devastated by a violent tornado. It had been a terrible year for storms—the tornado that hit that region was only one of 362 that appeared that year, a period that is now known as the 2011 Super Outbreak. It was the largest and costliest series of tornadoes ever recorded. "We had this spate of just one powerful tornado after another after another, in tornado alley and down into the south," she told me. "It was really horrendous and there was a lot of damage that was done. NOAA had done some great forecasts to get the word out, to warn people to get them out of harm's way." Still, there were 324 storm-related deaths related to the Super Outbreak.

Lubchenco visited the city in the aftermath of the storm with scientists who were evaluating the strength of the tornado and also to meet with representatives from FEMA, who were providing relief to the survivors. Lubchenco went on, "I didn't really realize this, because I hadn't thought about it too much, but with a tornado, there are no instruments on the ground, every place a tornado might be, to register wind speed." Unlike more regular storm or weather patterns, tornadoes are too focused, unpredictable, and violent to be consistently measured by conventional instruments while they are occurring. The only way to really measure a tornado's strength, she said, is to "go to a particular place where a tornado has touched down and look at the damage that was done to the buildings. You look at what kinds of buildings they were, how many stories, whether it was brick, whether it was wood, whether there's a foundation, whether the sides of the buildings are nailed to the foundation. If there are nails, you look at how far they are bent over when the superstructure is ripped off the foundation."

On this tour of the site, she was accompanied by a local politician, a fervent climate change denier. But as they both took in the material damage and spoke to the survivors of the storm, the politician, without prompting, suddenly wanted to talk to her about climate change. "He was very eloquent in saying 'this is now real to me. I'm getting this. I'm going to do what I can to help you all get the resources you need to help keep us safe. I've seen the light.'"

That last quote is worth repeating. A climate change denier, a lawmaker, who "has seen the light."*

---

* The irony, Lubchenco explained to me, was that she never suggested there was any connection between climate change and tornadoes. No serious climate scientist would connect those dots, she told me. Nevertheless, this intense experience with the destructive power of the weather had opened up the mind of a previously strong critic of climate science.

Lubchenco's anecdote reinforced my concern that unless an individual has a strong ability or inclination to imagine environmental damage, or has been directly affected by issues like climate change, he or she will likely be unwilling to make the hard choices needed to address this major challenge. Seeing is believing. This, no doubt, is part of the mystery of why, in spite of the fact that there is nearly universal agreement among scientists that climate change is real and an urgent threat to humanity, so many people remain skeptical. Sure, many people don't want to face the grim facts and huge sacrifices that would have to be made to address the problem—a fault we're all guilty of to differing degrees. Of course, many were being consciously misled by climate deniers' propaganda, but the more banal explanation is that people didn't believe it because they couldn't see it. Scientists have. Scientists are out in the field. They've been at the reefs and at the icecaps and they can see the clear and devastating evidence on their slides and core samples and pH meters.

Lubchenco's testimony is a reminder of the stakes involved in educating the public about climate change and effecting long-overdue political action. But it also reaffirmed my core belief that visceral effects of virtual reality could play a vital role in bringing these changes about.

ISCHIA

In those early years of using the lab to study environmental VR, I was struggling to get funding to support this new research interest. I wrote three separate large government grants to study how best to use VR to help people visualize and understand climate change, the same way the technology has been a success in other areas of science such as genetics and physics. Despite having had a stellar success rate with government funding—up until that point I had been batting over .500 on proposals funded—my climate change work was consistently turned down. One grant reviewer even hinted, dispiritingly, that I shouldn't be teaching school-children about climate change when it still had not been proven, and instead should "give them the scientific literacy they need to decide about climate change on their own." I decided to seek out help, and in 2012, found a partner in crime, my Stanford colleague Roy Pea, a pioneer in educational technology. I had known Roy since 1995—he was one of my professors in the Institute of Learning Science at Northwestern, the first school dedicated to studying learning technology. I like to joke that Roy has forgotten more about educational technology than anyone else will ever know, and he had a keen interest in using VR to help educate the world about climate change. It was a natural pairing.

He and I began submitting a series of proposals to science foundations to help fund a project that would combine Roy's expertise in using technology for education, with my lab's ability to create immersive VR. Initially we were focused on educating people about the great garbage patch, an enormous vortex of plastic waste, sludge, and other human-made refuse floating in the Pacific Ocean. The huge scale and remoteness of the garbage patch was just the kind of problem we felt VR could help educate the public about. But our proposal was rejected, and the grant organization suggested we team up with a marine scientist and sent us a list of possible collaborators.

That is how we became acquainted with the marine biologists Fiorenza Micheli and Kristy Kroeker and their work studying ocean acidification on a rocky reef in the shallow waters off the coast of Ischia, a small island on the western edge of Italy's Bay of Naples. Though I've spent a lot of time in its virtual waters, I've never actually been to Ischia—when your new line of research involves a resort island in the Mediterranean, it's important to signal to the grant organizations that it isn't a boondoggle. My loss was my graduate students' gain, and they reported back that Ischia was indeed a lovely place, with spas, bars, and restaurants dotting a lush landscape that culminates in the 2,500-foot peak of Mount Epomèo. The mild Mediterranean weather and stunning views of the Tyrrhenian Sea, however, barely hint at the powerful forces at work underneath this part of Italy. Mount Epomèo is an active volcano, a part of what's known as the Campanian Volcanic Arc, an area of intense volcanic and seismic activity caused by the meeting of the African and Eurasian continental plates. It is one of the most geologically volatile places on Earth, its dangers only heightened by the fact that 3 million people live in the vicinity, mostly in the densely populated area in and around Naples.

If you stand at the top of Mount Epomèo on a clear day, you can see, only a few miles to the east, the humpbacked shape of the most famous peak in the Campanian Arc, Mount Vesuvius. During its massive eruption in AD 79, Vesuvius dumped millions of tons of ash on the area, devastating the nearby Roman settlements of Herculaneum and Pompeii. In a few short hours, Pompeii was transformed into a mausoleum, one of the great cautionary tales of human history: beware the unpredictable wrath of Mother Nature.

Because of the volcanic activity in the area, Ischia is home to many hydrothermal vents, tiny fissures through which gases from deep in the Earth escape into the atmosphere. These fissures are responsible for the hot springs that make Ischia a particularly popular tourist destination for spa lovers. But they are also the cause of a rare feature off the northern and southern coasts of the island, one that has made Ischia a prized site for scientists concerned with the future health of our planet's oceans.

Vents such as these are extremely rare, and to find them so close to shore was a major discovery for marine scientists, who immediately saw their value in studying the effects on the oceans of a major greenhouse gas. In fact, the vents create a nearly perfect natural laboratory for this purpose. Their most noteworthy and scientifically useful feature is that the gas from the vents, which rises up in thin columns from the ocean floor like bubbles in champagne, contains almost pure CO_2 with virtually no accompanying hydrogen sulfide, a common by-product of volcanic vents (and the reason locations with volcanic venting often smell of sulphur). Because of the purity of the gas, whatever chemical reactions occur in the water around the vents are due to the effects of CO_2 and nothing else. Additionally, the vents are the same temperature as the surrounding water, meaning that any effect on the nearby plant and animal life was not due to an increase in heat that is common with volcanic vents.

The vents also created a natural gradient of CO_2 concentration that could be studied comparatively. And because the vents appeared to have existed for hundreds of years, existing plants and animals in the vicinity could be studied to see how they'd already adapted to the extreme conditions. For marine biologists interested in how increasing levels of CO_2 were affecting ocean life, Ischia was a model environment, a window into the future.

The reefs gave them the opportunity to analyze various organisms exposed to different concentrations of acidity, to see how each developed over time. Three distinct zones of acidity (ambient, low, and extreme low) were delineated after carefully monitoring the pH levels through sensors and repeated sample testing.

Their study had the benefit of not isolating one species, but instead of analyzing how many species interacted when dealing with the environmental stress.

OCEAN ACIDIFICATION

As Roy and I examined Fio's research about the vents, the clearer it became that the huge amounts of CO_2 we've been pumping into the atmosphere since the Industrial Revolution are leading to serious consequences for the plant and animal life in the oceans. When CO_2 hits the water, the pH level of the water goes down, and ocean acidification occurs. Higher acid concentrations interfere with the ability of many organisms to develop, including oysters, clams, lobsters, and corals. Terapods, a vital source of food for many fish, are also affected. Those that are better able to withstand the dramatic changes in the ocean chemistry, such as algae, thrive and choke out the life of these other groups. If oceans become too acidic, some believe, coral reefs as we know them will disappear altogether—and with them the rich diversity of life that they have made possible.

Scientists estimate one-third to one-half of all the extra CO_2 produced by humans over the past 200 years has been absorbed into the oceans. Our best estimates today tell us that the oceans are absorbing more CO_2 than ever before—about 25 million tons of human-made CO_2 *per day*. The result is that over the past two centuries the levels of acidity in the oceans have increased a staggering 25%, and the rate of change is accelerating. According to Thomas Lovejoy, the former chief biodiversity advisor to the World Bank, "The acidity of the oceans will more than double in the next 40 years. This rate is 100 times faster than any changes in ocean acidity in the last 20 million years."[11] It is unlikely, Lovejoy continues, "that marine life can somehow adapt to the changes."

Richard Feely, of NOAA, is also pessimistic, stating that it's probably already too late for the oceans to correct this damage naturally: "Fifty-five million years ago when we had an event like this (and that took over 10,000 years to occur), it took the oceans over 125,000 years to recover, just to get the chemistry back to normal. . . . It took two to 10 million years for the organisms to re-evolve, to get back into a normal situation. . . . So what we do over the next 100 years or 200 years can have implications for ocean ecosystems from tens of thousands to millions of years. That's the implication of what we're doing to the oceans right now."[12]

Ocean acidification has received less attention than it deserves. When I give public talks about using VR to educate people about the Ischia vents, I ask audience members to raise their hands if they've heard of ocean acidification. The usual response is less than one-tenth. It's not hard to understand why this is: under-water ocean environments are hard to visualize and difficult to reach, and the effects of acidification, at the moment, are mostly subtle. Most of us have no idea what a healthy ocean looks like, let alone one that is slowly dying due to the introduction of an invisible gas.

Instead, much of the study of climate change has been focused on the ways greenhouse gases directly affect life on land, which is why dramatic changes in terrestrial temperature, bizarre weather patterns, and the like receive much of the attention in the news media. Pictures of polar bears floating on melting ice floes, head-lines touting the fourth consecutive hottest winter/spring/summer on record, a red/orange heat map of global land temperature—these make much more dramatic news, and they literally scare us where we live. But the threat greenhouse gases are posing to our oceans is no less grave or starkly demonstrated by science.

These theories about the process by which the ocean becomes acidified, the effect that acidification has on plant and animal spe-cies, and the grave consequences these changes would have for all species tied to life in the shallow waters of our oceans, were confirmed by Kristy's and Fio's findings. As Fio puts it, "These reefs are like a crystal ball. We can understand future human impacts on the ocean by studying them."

In April 2013, Fio, Kristy, Roy, and I began to work out the initial design of the Ischia experience. Our goal was to educate users about the dangers of ocean acidification in an immersive simulation of the reef area that would be interactive, engaging, and scientifically accurate. We also wanted to create an experience that was as realistic and persuasive as possible. It was important to me that this experience feel real. In addition to the computer graphic recreation of the reef, we validated the simulation with real images acquired with a 360 degree video camera establishing the fact of these reefs, these vents, and the lack of biodiversity around them. If this experience were to change minds, people had to know that what we were portraying was true.

Kristy agreed with me, telling me a story about when her climate change–denying father visited her in Ischia while she was doing research. She'd grown up scuba diving along the Pacific Coast with her father, which was one of the reasons she became a marine biologist. But in spite of his love of the ocean, he was unconvinced by the science behind climate change. How could the actions of humans affect a planet that was so large? But when he and his daughter were able to dive together again, this time at the Ischia reefs, his mind was changed. Forced to see the deleterious effects of high CO_2 concentrations with his own eyes, he finally understood the work his daughter was doing and the stakes for the planet. He had read her academic writing, but after a direct experience he saw it in a different light.

Using 360 degree images brought back from Ischia, we built a computer-generated environment of the rocky reef area along with the columns of bubbles released by the vents. Great attention was paid to the modeling of the textures of the flora and fauna to create as realistic a visual simulation of the underwater environment as we could. Different animals and plants were animated and tediously placed in the proper locations in the demo. Fio and Kristy would then come by the lab to weigh in on the accuracy of the model. As we made corrections to these visual and content elements, Roy would suggest different interactive features, like a scavenger hunt that would encourage users to explore the different parts of the reef. Users would be asked to find, say, a certain kind of shellfish. As they searched the reef, they would notice that higher concentrations of the animal would be found in the areas away from the vents, thus illustrating the depletion of calciferous life in the high acidity areas.

We decided to incorporate the experience in the reef into a narrative that tied this remote location, and the invisible chemical processes of CO_2 pollution and ocean acidification, in to the fabric of our everyday life. The goal was for someone to come into VR and realize that their own behaviors were part of the problem. We also wanted to employ surreal elements, like the ones we used in the coal study, to heighten the effect of the experience.

In our narrative, users first find themselves behind a car on a generic city street. It could be their own hometown. After a few moments to get acclimated to the VR environment, the user begins to see CO_2 molecules spew forth from the car's exhaust pipe. The user follows one of the CO_2 molecules as it makes its way to the sea. There she uses her hand to push the CO_2 molecule into the water, and watches it combine with H_2O to form HCO_3, the acidic compound that is responsible for ocean acidification. Next she is taken on a field trip to Ischia to examine the shallow seafloor to observe, as if she is a scientist in scuba gear, the differences between a healthy reef and one severely damaged by acidification.

In our Ischia demo we had been exploring the many ways that VR can convey information about climate change: by being interactive, by altering time and physical scale, and by making invisible molecules appear throughout the air. All of these elements deepen and enhance the experience. It hasn't been easy—we've devoted thousands of hours to build the demo and it couldn't have happened without a $913,000 grant from the Gordon and Betty Moore Foundation. The coral in the Ischia demo feels real, and seeing the vibrant healthy coral ecosystem deteriorate around you brings home the message of climate change in a visceral and intimate way—far more powerful than a graph or chart. By creating convincing and scientifically accurate experiences of environmental destruction, perhaps the lessons scientists are trying to teach about the dangers can be more effectively conveyed. Fortunately, in VR, disaster is free. It happens at the touch of a button. Nobody gets hurt, but—just as in the pit—the brain responds as if it were real.

To date, we have shown this experience to thousands of people. It has been seen by US senators. An English prince. Thousands of students. "A List" Hollywood actors, producers, and directors. Professional athletes. We have uploaded it to STEAM, the "iTunes of VR," where it is downloaded by VR enthusiasts on a daily basis. Because of VR, ocean acidification is slowly gaining the attention of people who wouldn't otherwise know or care about it.

ECOTOURISM REVISITED

VR isn't just going to educate people about environmental damage: it's also going to make it possible to appreciate the beauty and majesty of nature in ways that don't impact natural habitats, and will arguably provide richer experiences than ecotourism or trips to zoos and aquariums. These possibilities were brought home to me by a whale-watching guide I met while on vacation with my family in Alaska in 2013. Blake was putting himself through school in a marine science program in Juneau by pointing out whales from the deck of a 25-foot boat. Whale-watching is one of the more popular tourist activities in Alaska. I was kept fairly busy corralling my two-year-old daughter, who was treating the boat as her personal playground, as I simultaneously tried to crane my neck around my family and the other ecotourists gathering on the railing. Then we saw it—a hint of a whale's back and a flip of its tail as it dove out of sight 300 feet away. It was amazing to witness this glorious animal in its natural environment—according to Blake we were lucky to see it. But I must confess, it was not as exciting as I had imagined. Blake told me that some people travel all the way to Alaska and don't see whales at all.

In virtual reality, there will always be plenty of whales (in fact, they are a bit of a trope in VR content), and tourists can see them as close up as they want, underwater or at the surface, swimming with a pod or alone. In fact, users can play the role of Jonah if they want, and walk around in the whale's belly. The weather is always perfect, the visibility always high, and the whales are engaging in whatever behavior is most educational—for example, any of the "Four Fs"—feeding, fighting, fleeing, or, um, mating. The holy grail of whale watchers—to see the animals jump out of the water and breach, slapping their bodies against the ocean's surface, would be a fairly simple animation sequence for a virtual programmer.

The best part of this is that virtual whales don't get harmed by tourism. Each year, according to our tour guide, more and more whale-watching boats were crowding the channel in Juneau. The tourism takes its toll on the animals. I asked Blake, who had one of the most sought-after jobs in Juneau (remember, he gets to look at whales for a living), what he thought of the idea of virtual whale-watching. He didn't hesitate. "I'd trade for virtual whale-watching in a second." He truly believed that our viewing the whales—despite government regulations and everyone's best intentions—was hurting the majestic animals. He also reminded us that the cruise ship we traveled on to get to this tour used about a gallon of gas for every foot it traveled.

Readers might at this point still rather see a real whale then swim with a virtual one. But it's important to consider scale: millions of people, flying to Alaska and then clambering into gas-powered boats. It takes a huge toll on the environment and the health of these species. How many more are invading natural habitats on safari in Africa? There needs to be some compromise.

As part of the regular outreach for my lab, we regularly bring in inner-city schoolchildren to experience virtual reality. Almost all kids love videogames, so coming to the most advanced virtual reality lab is a treat for them. One tour in particular illustrates how valuable virtual ecotourism—including new forms of zoos, aquariums, and group tours—can be.

In May 2013, a group of about two dozen sixth and seventh graders from Bridgeway Island Elementary School came to the lab. It's about a three-hour drive from Sacramento to the Monterey Bay Aquarium—one of the most highly regarded in the world. Despite this relatively short distance, many of them had never been to an aquarium before. More than a few had never seen a shark before.

One seventh grader got to become a shark. He put on the helmet and was transported to the bottom of a kelp forest, about 30 feet below the ocean's surface. After he'd had a few minutes to look around, seeing schools of bright yellow fish darting through the kelp, we asked him to raise his arms over his head. He did so, tentatively, and felt himself lurch upward in the virtual scene. As he moved his arms in the physical room, he learned to control the speed and direction of his virtual swim through the ocean. After a few moments he was shouting with delight. He'd never been snorkeling or scuba diving, and for readers who have not seen a kelp forest from underwater, it is truly majestic. After he had swum around for a few minutes, we asked the boy to find the shark.

We'd programmed the shark to be about 12 feet long and to swim on a random trajectory in the scene. But we did something that a live shark wouldn't appreciate very much—we allowed its skin to be passive. So a virtual swimmer could use his hands to match the speed and direction of the animal, and then literally enter its body to become the shark, and complete the ocean exploration *as* a fish. The boy shouted out, terrified when he saw the shark. But we assured him it was friendly, and coaxed him to try to swim inside its body. After some cajoling from his friends he successfully entered the shark and, from where we were sitting, had a blast. In just a few short minutes he had gained an appreciation for nature. On his way out the door he proclaimed he was going swimming in the ocean the following day.

CONSUMING MORE, PRODUCING LESS

Finally, it's worth considering the possible benefits of how our interactions with virtual worlds will affect our consumer behaviors and the waste produced by them. If we can reduce the number of trinkets and nonessential items that people, especially Americans, seem to be obsessed with buying, we will have won a huge battle for the environment. Factories require tons of natural resources to function, often pollute the environment during fabrication, and more often than not mass-produce plastic items that will likely be around longer than the human race. Pixels may require some energy to draw, but when the computer is turned off the output doesn't follow the path of the plastic bag, ending up in "The Great Pacific Garbage Patch."

It will take some time before we get used to this new economic paradigm. I recall a workshop I led for about 40 executives who worked with apparel on how virtual reality might transform their industry. One of them was skeptical when I said that kids in elementary school today might spend more money when they grow up on a virtual sweater than they would on one made of wool.

I told him about Veronica Brown. Brown, according to an article in the *Washington Post* from 2006, was "a hot fashion designer, making a living off the virtual lingerie and formalwear she sells inside the online fantasy world Second Life." The article went on to posit she would earn about $60,000 that year from selling her virtual clothing. That brought a few polite smiles and subtle snickers from the crowd, all of whom deal with numbers with far more zeroes than that on a daily basis. With the straw man officially set up, I moved on to my next data point.[13]

"How many of you have heard of the game Farmville?" I asked. Most of them raised their hands—after all this was the year 2013 and we were sipping coffee in the heart of Silicon Valley. "How many of you have played it?" One hand went up, tentatively. I put up the 2010 revenue figures from Zynga, the company that sells Farmville as well as other similar games. In a single year they had made over a half billion dollars. While that statistic might not have been new to all of them, everyone was blown away by what came next. Only a tiny fraction of that revenue came from advertising— almost every dollar came from the sale of virtual goods, including virtual food. At the time the company was selling 38,000 virtual items every second. In 2010, Zynga made $575 million in revenue off the sale of virtual goods, compared to $23 million from advertising. Selling virtual milk bottles to feed virtual baby animals is apparently lucrative.

"Do you know what they do in Farmville?" I asked. "They grow virtual food. Can you eat virtual food? Why do you think jackets and sneakers are any different?" These are extreme, and dated, examples, of course. Farmville and Second Life are no longer the objects of widespread cultural interest. But the behaviors people exhibited in those virtual worlds have merely moved on to other ones. Second Life and the thriving economies in virtual gaming worlds have shown us that people like to spend time and money in virtual worlds. Right now, users of all ages and backgrounds are spending real money to buy real estate, boats, and planes they can use in virtual spaces. They are spending money to customize their avatars with clothes, jewelry, and tattoos.

Massive economies and real-world wealth have been created in these worlds, which give users the pleasure of profligate consumption for relatively little cost. If it seems strange that people would shell out real money for avatar adornments and symbolic representations of status, then I encourage you to pause for a moment, look around you, and contemplate the pointless real-world consumer behavior that is a feature of our modern economy. The trouble is, conspicuous or wasteful consumption in the real word comes with real costs, be it in fossil fuel consumption, the mountains of plastic junk that are piling up in our homes and landfills, or the floating islands of garbage that are growing in our oceans, to name just a few. When considered this way, deep engagement in social virtual worlds seems less scary than the dystopian scenarios would have us believe, and may have significant social benefits.

CHAPTER 5

TIME MACHINES FOR TRAUMA

The patient, 26 years old, had stopped to buy something in a drugstore across the street from her office near the World Trade Center when the attack started on 9/11. She did not see the first plane hit the North Tower. In fact, few people did—how many people are staring up at the sky at 8:45 a.m. while on their way to the office? But the fire near the top of the tallest building in the city did attract interest, and so, after she exited the store, she stood with a crowd of onlookers and watched the smoke billow from high up in the tower, at the time seemingly the result of a tragic accident. That's when she saw the second plane fly into the tower.

It was this image, of the airplane cutting through the 77th and 85th floors of the South Tower just after 9:00 a.m., and the collapse of the buildings less than two hours later, that continued to haunt her several months after, as she and the rest of the traumatized city worked through the aftermath of the attack. Everything seemed to bring the horrors of that morning to mind: the constant television and newspaper coverage, the missing persons signs displayed throughout the city. The sight of the still-burning ruin in lower Manhattan, and the smell, which lingered for months as the fires burned out. Then there was the constant fear of another attack. Just standing in a canyon of a New York City street and looking up provoked anxiety for the patient. She was having trouble sleeping. She was prone to bouts of irritability and anger with her friends and family members. She couldn't stay at her boyfriend's apartment, which was located on a high floor in a tall building. Her family began to worry and finally sought professional help. The patient was "not herself," her mother told JoAnn Difede, a specialist in anxiety disorders at Cornell's Weill Medical College. After meeting the patient, Difede was quickly able to diagnose her as having classic symptoms of post-traumatic stress disorder.

Like many psychologists who specialize in treating patients with PTSD, Difede was seeing a surge of patients in the months after 9/11. Nearly 3,000 people died in the attacks in New York— a staggering number. But many thousands more experienced the events of that morning firsthand, inside the WTC buildings, on the streets around them, even stuck in subway cars underneath the massive complex. Thousands more watched from nearby buildings as the towers burned and fell. And of course there were the firefighters and police officers who rushed to the buildings to help. A decade after 9/11, of all those people who were exposed to that traumatic experience, at least 10,000 police officers, firefighters, and civilians were estimated to suffer from PTSD.[1]

As soon as she understood the size and scope of the attacks, Difede knew there would be thousands of people traumatized by the horrific events happening downtown and immediately began to prepare for a surge of patients. In the weeks that followed, she set up a screening protocol to find those exhibiting PTSD symptoms, which would eventually handle 3,900 individuals who had been exposed to the attacks. But once they were identified, how to treat them? At the time, the very idea of PTSD was still under debate—though it was being studied by many researchers, it had not yet been officially recognized by the DSM, the *Diagnostic and Statistical Manual of Mental Disorders*, as a psychological phenomenon.[2] Consequently, treatment options were limited. Many people suffering from symptoms were given antianxiety medications—a temporary fix that did not address the underlying issues. "It wasn't quite the Wild West," Difede told me, "but it was an emerging area."

Then and today, the most effective approach for treating PTSD was cognitive behavioral therapy (CBT) combined with imaginal exposure therapy. In this treatment, the therapist leads a patient through his memory of the traumatic event over the course of several sessions. Patients may be asked, for example, to close their eyes as they imagine the events and to write or recount what happened in the first person. The goal is to diminish the power of the trauma by forming a coherent memory of it. "You're creating a memory that's part of your life," Difede says. "It doesn't intrude when you don't want it to anymore."[3]

Crucial to the success of the treatment is that the patient's retelling not be a rote recitation of events—patients need to have an emotional connection to the events they are recounting, they need to reexperience them. One might go so far as to say they need to achieve a kind of presence. If these imaginative therapies go well, exposure therapy might lead a patient out of the office and back to the scene of the trauma—a process called in vivo exposure. The trouble is, reexperiencing a trauma means the patient has to overcome one of the primary symptoms of PTSD: avoidance. "We are wired to avoid pain," Difede explained to me at a VR conference in 2016, "whether it's trauma or something more mundane. And so it is a conundrum to try to confront your pain if avoidance is a coping strategy that happens preconsciously. So that's what you're up against with exposure therapy."[4] Difede realized that many patients were having trouble reconnecting with their memories, either because they were limited by their imaginative capacity, or because their traumatized minds weren't letting them access the painful memories.

This was not the first time Difede had experimented with VR. In the late 1990s, she had already begun investigating how to incorporate virtual reality into exposure therapy. "The idea of using VR was appealing," she told me, "because we know that our memories are not simply collections of verbal narratives. Our memories are typically sensory-rich experiences." An estimated 35–40% of patients do not respond to the standard, narrative-based imaginal therapy. Difede and a few other pioneers believed that a virtual setting filled with the sights, sounds—and even smells—that connect the patient's memory to the traumatic event being explored in the treatment might enhance outcomes. After 9/11, Difede saw an opportunity to test this theory on a large group of people suffering from acute PTSD. With some funding from the National Institutes of Health she reached out to a University of Washington psychologist, Hunter Hoffman, who had been using VR for phobia exposure therapy, and began a collaboration. In the frantic months that followed the attacks, Difede and Hoffman would speak almost daily as they developed the exposure model of downtown Manhattan, with Difede sharing the details gathered by her and her team from interviews with the survivors of the attacks. Then Hoffman's team in Seattle would code the computer-generated environment based on these interviews, incorporating sounds from news footage of the attacks.

Using a series of preprogrammed keystrokes, Difede was able to produce in the virtual world a progression of the events as they happened that day. Hitting one key on the keyboard, for example, would show the buildings before the attack. Another showed the burning North Tower in the minutes before the second one was struck. Another showed the WTC site after both towers had fallen. Using the keyboard, Difede could also include audio effects—of sirens, and screaming, and other sounds that survivors witnessed that day—allowing her to control the experience further. In this manner the therapist was something like a conductor of an orchestra, except that in this case she was managing the order and duration of virtual events in the way that was most effective for the therapeutic session.

The patient described in the chapter's opening would become the first person to use the VR environment Difede and Hoffman developed. Difede had already tried four sessions of imaginal therapy with the young woman before realizing that her patient wasn't responding. Her tone remained flat and affectless, and she clearly was unable to access the emotional memory of the attacks. So at the next session, after obtaining the patient's consent for the experimental treatment, Difede decided it was time to try the recently completed VR simulation of the attacks. Difede put the patient in an HMD and immersed her in the virtual New York. Now, as Difede guided her through her memory of that day, the patient could look around at the buildings and streets as she had on 9/11.

With the HMD on, the young woman was again standing outside that deli, looking at a simulation of those familiar city streets. The graphics were crude, but there was enough detail on the buildings and streets to put the patient in a recognizable location. Looking up she could see the towers rising above her—a familiar scene from her daily commute. She broke down in tears. She'd never thought she'd ever stand under those towers again. As the therapy continued over several sessions, Difede would talk her through her memory of the experience, and as they went over her narrative of that morning again and again, certain details returned to her. She remembered how she felt when the second plane hit, the horror when she realized this was not an accident. She remembered watching helplessly as the fires burned high above. She remembered, as the first tower began to fall, how she and the crowd she was with began to panic. In a journal article about the treatment, Difede described the patient's now vivid memory, facilitated by the virtual environment, as she tried to escape. She remembered being trapped under a "crush of bodies" as people fell over each other while they ran. After freeing herself she heard a cry and turned to meet the eyes of a woman begging for help. "She looked down to see that the woman's legs had been severed, and she was bleeding to death," Defide wrote in the case report. "Our patient recalled looking in the woman's eyes and telling the woman that she couldn't stop. Debris was falling all around them bringing along with it the possibility of a fatal collision. She recalled running and running through the hazy smoke."[5]

Eventually the patient remembered finding her way into a store further uptown. Her shoes had come off as she escaped the falling tower, and her feet were bleeding. She had no money. Looking around, she saw people seemingly going about their daily business. "Don't you know what's happening?" she screamed.

Defide believes the ability to recall these harrowing details was facilitated by the rich visual and aural environment of VR, which enabled the patient to better engage in the emotional content of her memory. As Difede explained to me, "We all layer our imagination and our memories into theatrical experiences, and there's a way in which a VR environment is a theatrical experience. And we bring meaning to our world, and this was an opportunity to do that." The diagnostic results for her first patient were impressive: Difede was able to report that after six sessions in VR, the patient showed an 83% reduction in symptoms of depression and a 90% reduction of symptoms of PTSD.

Difede's second patient, whom she started treating in VR in November of 2001, was a New York City Fire Department battalion chief. The firefighter had been in the command center located in the lobby of the North Tower at the moment the South Tower collapsed. Somehow, he survived. Covered with broken glass and debris, he was able to flee the area before the North Tower collapsed.[6]

Research has shown that people who receive acclimation training before entering dangerous situations—such as soldiers heading off to war—can be made somewhat more resistant to PTSD, but no amount of preparation can guarantee against its onset. This firefighter, for instance, was a military veteran, but none of his prior training was able to stop PTSD from taking hold. In the months after the attacks he experienced nightmares and had trouble sleeping. He feared enclosed spaces and, like Difede's previous patient, grew uncomfortable standing under tall buildings. He began avoiding Manhattan, and was taking Ambien to get to sleep. After his doctor refused to refill his prescription and recommended he see a psychologist, he finally sought out Difede and was introduced to her new, experimental treatment.

At weekly sessions, Difede would take the firefighter through a series of sequences from the day, reconstructing his viewpoint of the morning's events inside the virtual environment, while asking questions about what he was seeing and remembering. As with her other patients, Difede noticed his immediate, visceral reaction to the world. When confronted with the sights and sounds of that day, he would break out into a sweat and his heart would begin to race. And like the first patient, the emotional power of re-experiencing the event in VR put him in touch with memories he hadn't previously recalled, leading to a pivotal moment in his therapy.

At this point, months after the towers fell, the firefighter had grown accustomed to recounting his narrative verbally to reporters and investigators. Difede believes that the virtual scenario bypassed his conscious agency over how he told his story. Being present in the virtual world, she explained, primed his memory for the sensory elements and cues to fear in a way that just talking alone couldn't. By the fourth session he was recalling new details, including an encounter he had completely forgotten with a man in the vestibule of a building after the towers fell. The firefighter "saw himself running," Difede remembered. "The second tower had just collapsed and they were literally running for their lives. . . . As he was running he passed by an entryway and saw a guy in a blue jacket with FBI lettering on it, talking into his walkie-talkie." It was then, as he was escaping with clouds of dust and smoke behind him, that he recalled hearing the FBI agent say there was a third plane coming." (They were actually speaking of the planes bound for Washington, D.C.) At that moment, the firefighter suffered a panic attack at the news. He felt that he was going to die.[7] "That one piece of information completely changed his assessment of the situation," Difede told me. "One of the best predictors of PTSD is actually believing you're going to die."

Having found the specific moment of acute trauma that was the source of his anxiety, she and the firefighter could finally work on treating the memory. A few years after treatment, he was able to say that he "had his life back." Though he still experienced some anxiety around tall buildings and bridges, he was sleeping well again, his nightmares had stopped, and he was no longer on medication. "I can function and I couldn't have ever done that before this treatment," the firefighter told a journalist in 2005. "It's always going to be part of your life and there's always going to be something there that makes you different than you were four years ago, but now I'm a functioning person."

In the years after the attacks, Difede used her virtual reality model to treat over fifty patients suffering PTSD caused by the 9/11 events. This included a comparative study contrasting patients using VR with imaginal therapy and those using imaginal therapy alone. Confirming her anecdotal experience, she found the patients in VR demonstrated considerable statistical and clinical improvements in the outcome of treatment.[8] Since Difede's work with the WTC patients, other researchers have gone on to create therapeutic VR environments for a wide range of experiences from a terrorist bus-bombing in Israel to motor vehicle accidents. But the most extensive and pervasive use of VR in PTSD therapy has come in the treatment of combat veterans, by far the group most vulnerable to PTSD. Difede herself has begun a program to address combat-related PTSD, and in doing so she has teamed up with a researcher whose name has become synonymous with combat-related PTSD, Albert "Skip" Rizzo of the Institute for Creative Technologies at the University of Southern California.

I have always been jealous of Skip Rizzo. Not because of his brilliance, but because in 1974 he was in the crowd at the Ontario Speedway to see Black Sabbath play at the California Jam. I first met him in 2000 in Santa Barbara, where he presented his work to the virtual reality group at UCSB. He and I immediately hit it off, and not only because we both had long hair and liked heavy metal music. I appreciated the passion he had for his work, but also his easygoing attitude. Rizzo does not look like your stereotypical clinical research scientist. He loves motorcycles and he's an avid rugby player. His desk at USC's Institute for Creative Technologies is strewn with decorative skulls—an appropriate theme for someone who has dedicated his life to treating damaged brains. After training in clinical psychology and neuropsychology, he spent his early career studying PTSD and cognitive rehabilitation, developing rehab programs for people with brains injured by car accidents, strokes, and other traumas.

Skip has always been interested in how technology can improve treatments for his patients. He first started thinking along these lines in 1989. One of his patients at the time, a 22-year-old with a frontal lobe injury from a car accident, had been struggling to stay motivated and engaged in tasks for more than a few minutes at a time—a common symptom of damage to this part of the brain, which controls executive function. One day, before treatment, Skip observed the young man bent over, staring intently at a small screen. "What's that?" Skip asked the patient. "It's a Gameboy," the young man replied, and he proceeded to show Skip the famously addictive Russian puzzle game, Tetris. Skip watched for 10 minutes as the normally easily distractible frontal lobe patient sat glued to the game. "I thought, if only I could develop cognitive therapy that could engage folks like this." Immediately, he began incorporating games like SimCity into his clinical practice.

Not long after this experience Skip heard an interview on the radio with Jaron Lanier. Lanier was touting the work of his company, VPL Research, and the transformative possibilities of virtual reality. Immediately Skip saw the therapeutic potential of virtual environments for treating people with cognitive impairments and anxiety disorders: "I thought, what if we could immerse people in functionally relevant environments and do rehab in those contexts? Then we could build in a game element, some way to engage people." For Rizzo, VR could be the "ultimate Skinner Box," a controlled setting in which treatment approaches involving conditioning and training could be studied and carried out. Excited about these possibilities, he attended a conference in 1993 organized by Walter Greenleaf, who remains one of the most influential voices in medical VR to this day. Skip left this early demonstration of VR with two impressions: that VR was extremely expensive, and that little research was being done in the cognitive realm. He did see a study demonstrating how certain individuals with Down syndrome could learn how to navigate a real-world shopping trip by practicing in a virtual supermarket. The possibilities seemed endless.

Not long after that conference, he finally tried VR for the first time. Although he had immersed himself in the academic and popular literature on VR, and even published a paper speculating about its potential therapeutic uses, he had not yet had an opportunity to actually do a demo until that year. He was simultaneously thrilled and terribly underwhelmed. "At first," he told me, "I was like, 'Oh man, this sucks!' It was all geometric block buildings. The interface was hard to control. I got stuck in a wall because of bad collision detection. It was nothing at all like what I'd thought." But the potential was there. And after considering how quickly computers were advancing, he figured that by 2000 it would be in good shape.

Then came what Skip describes as the "nuclear winter" of VR, the post-hype Dark Ages of the late '90s, in which VR all but disappeared from the public imagination while researchers like Skip experimented in the cloisters of universities and corporate labs. At that point, Skip was using VR to treat Alzheimer's patients by doing innovative research in virtual object manipulation and developing training scenarios for kids with ADHD. He was also messing around with more consumer-based uses—some of the VR applications that are today touted as novel were being tried by Skip in those early years. By the early 2000s, he was experimenting with 360 immersive videos, which included the recording of a Duran Duran show at LA House of Blues as well as an exercise in immersive journalism, documenting homelessness in 360 video in Los Angeles's skid row. But his interests were always in his roots, cognitive rehabilitation and PTSD. And like JoAnn Difede, he became intrigued by VR's potential in helping his military patients during imaginal therapy.

In fact, the very earliest use of VR to treat PTSD patients involved Vietnam vets in the mid-'90s, in a treatment devised by Barbara Rothbaum at Emory University. Building on that work, and the success that Difede had had with 9/11 patients, Rizzo designed a system that would help with the huge influx of PTSD sufferers who were being created by the wars in Afghanistan and Iraq. Using the graphics engine of a popular first-person shooter combat game called *Full Spectrum Warrior*, Skip created a "Virtual Iraq" that allowed him to re-create scenes where soldiers experienced trauma throughout that theatre of war. Marketplaces, apartment buildings, mosques—any place where soldiers had experienced a gun battle, a suicide bomber, or an IED could be reconstructed, including sounds and even smells to deepen the sensory engagement with the simulation. Soldiers on foot patrol carry a properly weighted assault rifle with the movement controls mounted on the handguard to further create a sense of presence. It is extremely customizable. The therapist can choose from a large menu of effects and scenarios to re-create the individual experiences of patients, from the time of day to the sounds of voices in English or Arabic.

Today this program is known as BRAVEMIND. Available since 2004 in over 75 sites across the country, it is estimated to have helped over 2,000 soldiers suffering from PTSD—particularly younger ones who might be less comfortable engaging in verbal imaginal therapy and are more at ease interacting with digital worlds. Using technology to recapture the experience was natural to them, and they were already familiar with violent war videogames. Rizzo, however, is quick to point out the difference between a system like BRAVEMIND and a game. "We're not throwing someone into *Call of Duty*," he says, referencing the popular war-themed videogame. "They do that anyway—but that's a cathartic revenge fantasy. What we do is try to induce anxiety in people by having a clinician, using a control panel, adjust the time of day, the weather, the lighting, the sound effects, the explosions, and help the patient confront the process, everything they've been avoiding. So, it's not a game. It's using game technology."

Like many of the people who have worked with and thought about VR for decades, Rizzo is overflowing with practical applications for the technology to facilitate healing. Among them are a PTSD application for people who have suffered from sexual assault and a virtual classroom to help kids with ADHD. He's built a prototype of a virtual therapist to help soldiers understand PTSD and ameliorate the shortfall in clinical therapists. He's created virtual patients to help train psychologists and social workers about how to interact with people suffering from mental illness. Medical errors in hospitals, he points out, cause 38,000 deaths a year. Virtual training for doctors and nurses could reduce that number.

In spite of the success Difede and Rizzo have had with VR exposure therapy, and the generous support they have received from a military that is eager to address the huge numbers of PTSD patients created by the wars in Afghanistan and Iraq, there still

remains resistance from some quarters of the psychological estab-
lishment. Many therapists worry about the intensity of VR, and
the dangers of exposing patients with fragile psyches to a flood
of painful memories. Difede points out that the cues that trigger
anxiety which appear in daily life need to be addressed. As she told
The New Yorker, "If you suddenly become afraid of the staircase
because you walked down 25 flights of stairs to get out of the World
Trade Center, the stairs went from being neutral to negative."

Difede recognizes that widespread acceptance can take time.
"My generation and older were not trained to think this way. There
is a cultural shift in these treatments." Resistance to change can
be a feature in medical innovation, of course, protecting patients
from poorly conceived and unproven remedies. But the evidence
for VR Immersive Therapy has been strong for some time now.
Still, it will take time before these treatments are widely adopted.
Difede points to an analysis she did on the adoption of new medical
technologies in 2014. "It takes 17 years for new research to go from
bench to bedside and be implemented in routine clinical practice.
So, in that context, it's not surprising. Has it got better, yes I think
it's gotten better. Is it where it should be? No. Absolutely not."

In people with PTSD, we can use VR to bring them closer to
reality, to heighten their emotions by programming real-seeming
environments and put them in touch with their memories. But this
isn't the only medically useful application of VR. The other major
one rests on what many consider to be the medium's most alarming characteristic—its ability to so engross us and distract us that
we lose connection with the real world.

CHAPTER 6

ABSENCE MAKES THE PAIN GROW FAINTER

n 2014, at a party with some colleagues from Stanford, I joined the ranks of millions of Americans who have suffered from lower back pain. It all started when my three-year-old daughter toddled over too close to the edge of a pool. Fearing she might fall in, with my paternal instincts kicking in, I lunged to intercept her. At the time I barely felt the slight twinge in my back, and dismissed it as nothing to worry about. An hour later I was lying on the floor of my host's back porch while professors politely stepped around me, staring at the ceiling and wondering how I would ever get up. I had never been in so much pain in my life.

I won't soon forget how excruciating daily life was in the months after I injured myself. Not a moment went by when I was not experiencing a general discomfort, occasionally punctuated by stabs of searing pain, as I attempted the most routine daily activities, like getting up from a chair or holding the daredevil child who started it all. Most readers will have some idea of what I'm talking about: 80% of adults in America will experience acute lower back pain in their lifetime, and at any given time, a quarter of Americans will have suffered its effects within the last three months. Lower back pain is so common, and so debilitating when it is occurring, that one study showed it to be the third-most burdensome condition in the United States, exceeded only by ischemic heart disease and chronic obstructive pulmonary disease.[1]

In the end, I was lucky—after six months of physical therapy exercises I was almost back to normal. But for many people, the pain doesn't go away—about 10% of the millions of people who suffer from lower back pain will find no relief through therapy or surgery. When such suffering persists beyond six months, medical health professionals categorize the pain as chronic. For these patients, the seemingly innocuous back tweak is a passport to a land of unremitting physical and psychological torment. An estimated 20–30% of Americans, or as many as 100 million people, suffer from chronic pain. While the origin of chronic pain can often be easily diagnosed, as in the case of lower back injury, its causes can also be complicated and mysterious, with no clear path of treatment. Either way, the grueling experience of dealing with constant pain carries over into every aspect of an individual's life, affecting everything from sleep, work productivity, and personal relationships to the patient's own mental health, and contributing to a negative feedback loop that can lead to severe depression.[2]

One way doctors have sought to treat those suffering from acute and chronic pain is to administer powerful prescription opioid painkillers like oxycodone and hydrocodone. Indeed, my doctor and I discussed this possibility before we decided to see first how I responded to physical therapy. He was no doubt wary due to a growing awareness in the medical community that opioid painkillers— as effective and necessary as they sometimes are—have been overprescribed in recent decades, and with grave unintended consequences. The surge in opioid use began in the mid-1990s, when lower prices and aggressive marketing from pharmaceutical companies led to a sharp increase in the prescription and use of these drugs. The results have been nothing less than catastrophic, as the "opioid epidemic," as it is now known, has ravaged countries around the globe, and in particular the United States, where deaths among heroin and prescription drug abusers top 27,000 a year. There were nearly 19,000 deaths from opioids alone in 2014, a 369% increase from 1999. In the same period, heroin overdoses increased 439%. A study done in 2014 revealed the remarkable fact that 12 US states had more opioid prescriptions than people.[3]

By 2010, the flood of opioids into the health care system and rising rates of addiction had grown so alarming that regulations were tightened to control "doctor shopping," "pill mills," and the widespread abuse. But these moves had a few unintended consequences. Demand for prescription painkillers in illegal markets skyrocketed, with an individual pill in 2014 costing as much as $80. This drove new addicts to street heroin, which could be obtained for as little as $10.[4]

A new pattern of abuse began to emerge. A patient would experience an injury—say to the lower back—that would lead to an onset of chronic pain requiring surgery. In the aftermath of surgery, painkillers would be prescribed to help the patient cope with the acute pain. After the prescription runs out, the patient, still experiencing agonizing pain, or having developed a dependency on the pills, is suddenly cut off. With no graduated pain treatment plan or addiction resources available, a patient begins to treat herself with over-the-counter medications but finds little relief. Perhaps she borrows pain meds from a friend who has some left over from a previous surgery. Eventually she seeks pain remedies through illegal drug dealers, at first only sniffing heroin and then eventually injecting.

Looming behind these alarming numbers about chronic pain

and opioid use is another foreboding statistic. Health care experts anticipate the numbers of people experiencing the agonies of acute and chronic pain will increase in the coming decades, as the baby-boom generation continues to age and will live longer than any other generation in history. How we will safely treat a growing population suffering from chronic pain is an urgent problem that we as a society must address. One of the people looking into it is Sean Mackay, the head of the Pain Management Center at Stanford University Medical Center. A self-described "recovering anesthesiologist," Mackey has become a leading figure in efforts to understand and combat pain.

In a study published in 2016, he and his colleagues demonstrated the link between chronic opioid use and eleven common surgeries, including knee surgery, gallbladder surgery, and most alarmingly because of their frequency, cesarean sections.[5] Mackay points out that the crucial time to intervene, if we want to effectively treat the traumatic pain that accompanies surgery, is the moment immediately after surgery, when painkillers are first administered. He and a growing chorus of public health experts agree: we need to find new, nonpharmaceutical ways to relieve people of pain that eliminate or reduce the use of opioids.

There are a variety of these interventions already being used in the medical world, including techniques like massage, meditation, acupuncture, or pet therapy. Another important technique is distraction. Doctors and therapists will encourage patients suffering from pain to employ all manner of distraction techniques, from reading books and watching TV, to playing videogames and drawing. The reason distraction works is that human attention is finite. We can only attend to so many stimuli at once. And few things are more distracting than a medium that engulfs the senses and allows users to have custom-made experiences. The power of presence—the mental transportation into a virtual world—has a useful side effect: absence. Being present in a virtual world takes one's attention away from his own body.

As evidence mounts for the benefits of VR distraction from pain, we increasingly see VR being utilized around the world to manage physical pain of all sorts, reducing the discomfort and anxiety of people who flinch at needles and dental drills, and helping those who must undergo tedious and uncomfortable rehabilitation exercises to perform these tasks better. The original research that has led to the adoption of VR distraction into pain management began, as so many VR stories do, during the VR boom of the early '90s. And it involves the work of a researcher we've already met—JoAnn Difede's collaborator on her 9/11 PTSD work, Hunter Hoffman.

It all began with the kind of casual, collegial conversation that reminds us of the importance of universities in stimulating interdisciplinary collaboration and innovation. It was in 1996, and Hunter Hoffman was working at the University of Washington's recently launched Human Interface with Technology Lab (HIT-Lab), exploring how VR could be used for people suffering from phobias. Although he had begun his career studying memory and cognition, after doing a VR demo in the early '90s he became fascinated with the illusion of presence and began focusing on virtual reality. Using a prebuilt virtual environment and the university's VR equipment, he began a series of experiments on patients with arachnophobia, seeing if exposure to a virtual spider could help them cope with their fear. The results were promising.

One day, one of his friends in the university's psych department was describing to Hoffman how hypnosis therapy was being used as a pain control technique for burn patients. "I asked him," Hoffman recalled, "'how does hypnosis reduce pain?' and my friend said, 'Well, we don't fully understand how it works, but it may have something to do with distraction.' And I said, 'Oh my gosh, well I have an amazing distraction that will knock your socks off.'"

His friend put Hoffman in touch with David Patterson, a UW professor of psychology who specializes in rehabilitation psychology and pain control. Together they began looking into the effectiveness of VR in relieving victims of severe burns from the intense pain that was an unavoidable part of their treatment. The Harborview Burn Center in Seattle, Washington, where they conducted this research, is a regional hospital that handles patients from five neighboring states. Burn injuries, often involving wounds over large parts of the body, are known for their severity and complexity, and the course of treatment is uniquely painful.

First, skin for grafting onto the burned areas must be harvested off of unaffected areas on the body, introducing new wounds to torment the patient. The state of perpetual pain from the new and old wounds is then punctuated by constant treatment to facilitate healing. Each day, bandages must be removed and replaced, tearing off scabs. Then the raw, damaged skin is soaped and scrubbed to ward off infection and encourage the grafts to take. As the grafts begin connecting with the damaged skin, painful exercises must be performed to break up the developing scar tissue, so that the patient can recover movement. On the 1–10 pain scale that doctors use to evaluate patients' experiences of pain, these treatments routinely get the very highest rating, even with the help of opioid painkillers.

At the time Hoffman was conducting his experiment, opioids were the primary way to handle this excruciating experience. But while opioids alone were adequate at controlling pain when

patients were resting, they were largely ineffective during the daily
wound care. At those moments, nearly 90% of patients would report
severe to excruciating pain, even when on the medicine.[6] Even
with prescription painkillers, recovering from burn wounds is a
uniquely painful process, and offers a special challenge to doctors
and nurses attempting to provide relief to a patient. Painkillers,
as effective as they are at numbing the sensation of pain, can have
adverse effects on recovery. Not only are they addictive, but they
interfere with sleep, induce nausea, and, if improperly administered, can cause death. The benefits of distraction were already
well known to researchers—for years scientists studying pain had
tested the effectiveness of media like film, music, and videogames
at distracting the minds of patients during painful treatments. To
Hoffman, VR seemed like a logical next step. He suspected it would
be an effective distraction tool, but would it work better than other
distraction techniques? He and Patterson decided to test this idea.

In one of the early pilot studies by Hoffman and his colleagues,
the scientists employed two different distraction methods during the treatment of two burn patients. One involved immersion
in Hoffman's already built arachnophobia environment. "Spider-
World," as Hoffman called it, was basically a virtual kitchen, featuring countertops, a window, and cabinets that could be opened.
The star of the show was a furry Guyana bird–eating tarantula that
rested, alarmingly, on the countertop. Brave patients could even
touch the spider if they wanted to—Hoffman placed a physical representation of the tarantula, a "furry toy spider with a bad toupee,"
just within reach.[7] SpiderWorld was quite elaborate for the year
2000. In it you could grab plates, toasters, plants, and frying pans
using a virtual hand that was controlled by a glove. Users could
also rotate their heads and translate (i.e., move) their torso to lean
in different directions. In many ways the technology was similar to

the commercial systems that have become available to consumers today, though of course lower in fidelity and higher in cost.

The other distraction technique was a popular racing video-game made for the Nintendo 64 console, in which patients used a joystick to race and maneuver their race car or jet ski around a course. While the videogame was clearly not as immersive as the VR world, it was chosen by the authors because it was elaborate and extremely absorbing. Unlike SpiderWorld, it had scoreboards where patients got feedback about their performance, and was more involved in terms of the first-person task and narrative. In this way the control condition was not a "straw man"; it was not immersive VR but it was very engaging and distracting.

The first patient was a teenage boy, 16 years old, with severe burns on one leg that required surgery and the insertion of a sta-ple. The results were stunning. One metric that doctors use to quantify pain is to determine what percentage of the time, dur-ing a procedure, the patient is thinking about pain. In this study, during wound care, the patient thought about pain for 95% of the time while playing Nintendo, but only 2% of the time while in VR. Hoffman and colleagues demonstrated similarly large differences in ratings of unpleasantness, pain, and anxiety. VR had a massive benefit, and reduced pain dramatically compared to videogame play. The second patient had severe burns all over his body—his face, chest, back, stomach, legs, and right arm. The burns covered one-third of his total body surface area. His results were similar to the first patient's—VR reduced pain by huge percentages com-pared to the game. The article Hoffman and Patterson wrote on this study made the cover of the medical journal *Pain* in 2000, with a picture of a patient in an HMD. This was a surprise because *Pain* usually published articles about pain research with animal subjects, focused on the cellular level. "I think it was one of the first humans to be on the cover," Hoffman joked, "it's usually a picture of iridescent rat spines."

Needless to say, Hoffman and his colleagues were more than a little encouraged by these results. Not only was the VR environment extraordinarily distracting to the patient, but they had achieved these results with what amounted to an off-the-shelf experience. The "SpiderWorld" they had used—which depicted a kitchen filled with potentially unpleasant associations like an oven, a stovetop, and a toaster—was not exactly welcoming to burn patients. Hoffman wondered if he could create a more pleasant experience. He also wanted to turn the experience into a game—imagine how much more involving the treatment would be if it combined game design with the immersive properties of VR.

The result of Hoffman's tinkering was SnowWorld, a simple, sedate VR game set in a world of cool whites and blues. In it, the player/patient gently moves along an arctic canyon floor amidst falling snowflakes, snowmen, penguins, and woolly mammoths. Using a mouse, the patient can aim snowballs at the virtual objects, and defend herself from snowballs lobbed at her. All the while, a pleasant pop music soundtrack plays. It was important to Hoffman that the game be simple to play, and not too stimulating. "I designed SnowWorld so you move slowly through the environment," Hoffman told me. "We were worried about simulator sickness because people in the burn ward are already nauseated from their burn meds, as well as from their wounds. We really bent over backward to minimize simulator sickness, and what that meant was the patient follows a pre-defined path through the virtual world, and that really calms things down."

Four years after his original landmark study with the two patients—which remains one of the most referenced articles on VR and pain—Hoffman and colleagues used SnowWorld in research designed to study a different metric of pain-relief effectiveness.[8] While patient self-report ratings are the typical way to measure pain, it is also possible to look at patterns of brain activity to determine painfulness. To increase the sample size from two patients, they used nonclinical participants who did not have a preexisting condition. So that the variables could be better controlled, the healthy participants came to the lab and the researchers induced pain. Eight men participated. They entered an fMRI machine, and then held on to a special device called a thermode that slowly heats up until it causes a painful burning sensation in the palm of the hand. We have used one of these in my lab for a dissertation on VR and pain, and I can personally vouch for what an unpleasant experience it is. I actually hold the lab's record for the lowest pain tolerance in the group, and can only hold on to the thermode for a few seconds. Each participant had two trials, one with VR—they used the famous SnowWorld simulation—and one without VR. Half the subjects received the VR trial first, while the other half received the control condition first. Each of the two trials lasted about three minutes.

The scientists isolated five brain areas where they believed more activity would signal more pain, for example in the thalamus. They found less activation, according to the fMRI, in all five brain areas when the participants were in VR compared to the control condition. This was the first evidence to show that VR actually changes brain activity during painful procedures.[9]

Hoffman knew that in order to make VR pain treatments widespread he would have to convince hospitals and insurance companies to recognize VR's effectiveness. So he and his colleagues, in 2011, took a critical step toward gaining credibility in clinical fields: a randomized controlled trial.[10] These studies have larger samples and take great care to ensure the validity of the treatment and control conditions. Fifty-four hospitalized children—burn patients—participated in the study while doing physical therapy to treat the burns, which is a horribly painful experience in which the patients extend their range of motion by doing exercises. In each session, the patients spent about half of the time in VR, and the other half in the control condition, with careful attention to proper randomization to avoid order effects. During each session, subjects spent equivalent time in both the virtual reality and the control conditions (treatment order randomized and counterbalanced). The children experienced less pain when they were in VR, with decreases in pain ranging from 27% to 44% compared to the control condition. Moreover, they reported having more "fun" in VR compared to the control condition.[11] Fun is in quotes because the physical therapy is no walk in the park, and any motivation to increase compliance with a painful exercise regimen is a huge boon.

Hoffman's pain research is being applied to the very extremities of human suffering—few individuals have to endure as much acute pain as burn patients. But the potential applications for VR as a pain treatment are boundless and starting to appear in even the most common medical settings. Already, many doctors are using immersive video to help patients undergoing routine medical procedures like IV insertions and dental cleanings, and to reduce anxiety by transporting bed-bound patients to other places.[12] Doctors have used VR to help patients undergoing chemotherapy, who have reported that it made the duration of their treatment seem shorter.[13]

While applications for creating distraction have been the major area of research in VR pain management, new avenues are being explored that build upon some fascinating recent research into the complex interactions of the mind and body, including how creating the illusion of movement in an affected or missing limb can rewire the brain, which can encourage motion (when possible) and relieve pain. These treatments are based on mirror therapy, a technique originally devised in the 1990s by V. S. Ramachandran at the University of California–San Diego, to help people suffering from phantom limb pain, the uncomfortable sensations ranging from itching to burning which can affect people who have lost limbs. As many as 70% of amputees suffer from it.[14] The cause of phantom limb pain is still under debate, but one theory is that the somatosensory cortex in the brain is rewired upon the loss of nerve inputs from the missing limb, and this rewiring creates the painful sensations. Neural pathways stuck in painful loops—say, a phantom hand balled into a tight fist—can thus be relieved by tricking the brain into perceiving the phantom fist as being relaxed. This is done by using a mirror to reflect a healthy, unaffected limb in the place where the affected or missing limb was. With this technique doctors are able to create the illusion of a healthy limb, but one that can be exercised or moved in a way that relieves the pain. Patients would look in the mirror, and see their good arm reflected in the location where their amputated arm was supposed to be. With the proper exercises, the phantom fist could be relaxed, and the brain rewired.

A significant amount of research has shown mirror therapy to be remarkably effective for certain patients. But it still doesn't work for about 40% of those who try it. One theory is that a success is related to the ability of the subject to imagine the reflection of the limb as one's own.[15] Some people seem to have a harder time doing this than others. Because VR can replace a lot of this imaginative work, many believe it provides an enhanced version of mirror therapy, given how compelling the body transfer illusion can be.

My colleague Kim Bullock, a psychologist and physiologist, has recently been incorporating virtual reality therapy into her practice in treating people suffering from sensory, motor, and cognitive impairments. For two years we had been working on a project that uses avatar limb movement to treat pain, and had received a small grant from Stanford to test VR on different types of medical treatments. One of her patients, Carol P, had been having extremely good results from her VR therapy.

Carol P suffers from cerebral palsy and dystonia, which causes uncontrollable muscle contractions on the right side of her body, and confines her for all but brief periods in a wheelchair. Over time, those contractions led to the dislocation of her right arm from the shoulder, forcing her to rest the limb on her chest. She described the pain to me as constant, a throbbing sensation of "bone rubbing against bone," punctuated by sharp pains as if she were being stuck with a pin. Therefore, Carol tries to avoid moving her right arm as much as possible. But this is only the most acute aspect of Carol's pain, which also involves chronic general discomfort in the joints, cartilage, and ligaments throughout her body. Carol has lived with pain for her entire life, but after the shoulder dislocation it became difficult to go about her daily activities. In 2011, it became unbearable, so she sought help from doctors, who prescribed medication, massage, and deep brain stimulation, to limited effect. Still looking for relief, Carol was put in touch with Bullock.

The therapy works like this: Carol will put on an HMD and see herself in the first-person view via an avatar body. In her unaffected left physical arm, Carol will hold a motion-tracked controller of her avatar's arm. But instead of moving her virtual left arm, Bullock will switch the inputs so that Carol's real-world movements control her virtual right arm. She now has the illusion in VR of moving the dislocated arm that is painfully resting on her chest. She uses her left physical arm to move her right virtual one. But since her brain has taken ownership of the virtual body, it perceives itself as moving the dislocated arm.

After only a few treatments, Carol was reporting incredible results in pain reduction. When I spoke with her on the phone, she couldn't have been more enthusiastic about the gains she was making by exercising her affected arm. "Each session I went to," she told me excitedly, "it helped to activate it more. Up, down, sideways—I can turn my wrist any way, and what I noticed the first time was, that the actual sensation of feeling relief, was release from the pain at first, then it was just a sensation of knowing that it was coming out of its shell. I'm thrilled. I can move my arm and give it some exercise."

Carol was an inspiration, and it was wonderful to see how her virtual experiences were making her life better. And it wasn't just for the pain. Carol also felt liberated by the opportunities for novel experiences that VR offered her, as someone who has difficulty moving in the world. I asked her whether she'd had the opportunity to sample any VR apps not related to her therapy. Even over the phone, I could perceive her delight.

"It was amazing!" Carol told me. "I was in the ocean." Bullock had shown her a sample of a demo created for the Vive HMD, which places the user on the deck of a sunken ship. Fish swim about. If you look over the railing you can gaze into the abyss of the ocean below. Suddenly, a whale swims by the boat, its eye only a few feet from you. "It was like feeling free," Carol went on. "From not being in the chair. It was a wonderful feeling. Of having that independence."

"That was totally amazing. . . . And what I liked about it was, I'd never been scuba diving, it was an awesome feeling of knowing I was in the water with the fish and of having freedom."

I asked Carol what other experiences she'd like to try: "Skiing. Flying. Maybe. I'm not really sure, but that was the first time I went scuba diving and I loved it."

Mapping one's brain onto a virtual body, as Carol P does in her therapy, is producing some intriguing and effective clinical results. But the process raises some interesting theoretical questions. We know humans are very good at controlling their typical bipedal housing, but what happens when the avatars represent a completely different body form? Our brains have adapted so that we can effortlessly control our unique bodies, allowing us to turn our head, or move our arms and legs, with ease. We can diagram the brain to show which parts of the motor cortex control which parts of the body—a so-called "homunculus," an idea that references the ancient belief that a little person located in our heads controls our bodies, like a pilot in a cockpit. If you've seen these diagrams you'll have noticed that a large part of our cortical real estate is dedicated to our hands and face. But imagine if the homunculus is suddenly presented with an entirely new body schema: How do we cope when we are suddenly operating a body that isn't bipedal, say an avatar lobster or octopus? Can the brain adapt to figure out how to control six extra arms? The theory that examines this question is called Homuncular Flexiblity, and it is the brainchild of VR pioneer Jaron Lanier.

In the 1980s, Ann Lasko, who worked with Jaron at VPL, had seen a postcard picture of people in lobster suits at a festival. This inspired her to create a lobster avatar in VR, and she set about programming a body map for it. Since the lobster body includes six more limbs than a typical person, there were not enough parameters measured by VPL's body tracking suit to drive the lobster avatar in a one-to-one mapping. Therefore, the scientists had to creatively come up with extra mappings to extend the functionality of the body suit to the greater number of degrees of freedom of the lobster. For example, one could measure the movement of the physical left arm. That movement would of course directly control the movement of one of the lobster's arms, but simultaneously one could mathematically transform that physical movement—for example, repurpose a bicep-flex that wasn't essential to the position of that arm—and transform the flex to then control the position of a second virtual arm.

The initial mappings were not usable, but over time, as the algorithms to control the extra limbs evolved, some mappings emerged that succeeded. According to Jaron's observations, over time, and with practice, humans were able to slowly learn to control the lobster. The biologist Jim Bower, when visiting Jaron's lab during this period, commented that the range of usable nonhuman avatars might be related to the phylogenetic tree, an evolutionary memory. The human brain might be expected to find body plans that had occurred in the history of its own descent to be more usable than other body plans, though of course this would not include the lobster. Nonetheless, the intriguing question remains of what makes certain nonhuman avatars usable while others are not.[16]

One of my graduate students who runs a VR lab at Cornell University, Andrea Stevenson Won, was the first scientist brave enough to build, test, and publish a paper on Homuncular Flexibility. She and I worked with Jaron, who coauthored the eventual paper that reported our findings, to build the algorithms and metrics for testing body adaptation. In two studies, we manipulated one's ability to control novel bodies, and examined how experimental participants adapted to the new bodies.

The first study kept a human bipedal form, but shook up the control schemes for a normal body. We switched the tracking data of virtual arms and legs, such that virtual arms controlled the physical legs, and vice versa. There were three conditions: normal, switched source, and switched range. In the normal condition, when a participant arm moved her physical arms and legs, they were tracked and the avatar's limbs moved accordingly, with similar ranges. In the switched source condition, when the participant moved her physical legs, her avatar moved its virtual arms. Similarly, when the participant moved her physical arms, her avatar moved its virtual legs. In the extended range condition, the participant's arms and legs moved the appropriate avatar limbs, but the range of the avatar's limb movement was either expanded (in the case of the virtual legs, which now had the range of real-life arms) or contracted (in the case of the virtual arms, which now had the range of real-life legs, and didn't rise above the shoulder). In other words, in this condition the avatar was able to move its legs with more range than humans, and its arms with less range than humans.

Participants were tasked with popping balloons for about ten minutes. The balloons appeared in random spots in front of them and were reachable by either the virtual hands or feet. We were able to track the number of balloons they could pop with specific limbs, as well as the overall movement of the four limbs. Results demonstrated adaptation. After about four minutes, on average, participants would stop being disoriented and would learn, as measured by how many balloons they could pop, how to accommodate to the new body structures. In the switched condition and the extended condition, they would move their physical legs more, even though people avoided leg movements in the normal condition. People adapted to the strange virtual bodies fairly quickly.[17]

After this first theoretical study, which was designed to learn about brain mapping, we actually found a company that wanted to fund the work. A Japanese corporation called NEC wanted to explore how to make workers more productive. Well, what if you had a third arm? Would this make us more efficient workers in factory lines, data clouds, and all facets of life? One could make the argument that a third arm would be distracting, and that multitasking would lead to decreases in productivity. On the other hand, if one could master a third arm then it would be game-changing. So in this next study we gave avatars a third arm, which emanated from the middle of their chest, and extended about three meters forward. It was a very long arm. The participants could rotate their left arm at the shoulder to control the x position of their third arm, and could rotate their right arm at the shoulder to control the y position of their third arm. The arm rotations operated independently of the position of the left and right arm, so the control scheme did not interfere with the ability of the natural arms to perform a task. In this study, participants had to touch cubes that were floating in space. For those in the two-arm (i.e., normal) condition, some of the cubes were within arm's length of their body, while others required taking a step to touch as they were three meters in front of them. For those in the three-arm condition, the close cubes were reachable by the natural arms, and the far cubes were reachable by the third arm without walking. A trial was defined by completing a task of popping two close cubes and one far cube, all of which appeared at one time and then changed colors once the trial was complete. Participants took less than five minutes on average to master using the third arm. After five minutes, those with three arms routinely outperformed those with two arms. Having a third arm makes you more productive. Our corporate sponsors were delighted with the results.

An interesting tidbit is that Jaron stumbled upon Homuncular Flexibility as an accident, like many important scientific discoveries. In his early work in the 1980s, he and his colleagues were the first to build networked VR, at his company VPL Research. To be networked, users would experience being present in a shared virtual setting and would be able to see representations of each other. Thus it was necessary to create three-dimensional avatars of each user. One of the strengths of the early development system was that it supported extremely rapid prototyping, with revisions appearing instantly while a subject was "inside" a virtual world experiment. The calibration of early full body avatars presented a challenge; it was difficult to design a suit so that sensors would remain in precisely the same locations on the user's body with extended use. In the context of rapid experimentation with mappings and heuristics, there were, of course, bugs. These would usually result in a complete breakdown of usability. As an example, if an avatar's head were made to jut out of the side of the hip, the world would appear rotated awkwardly to the user, who would immediately become disoriented, unable to perform any task. In the course of exploring avatar designs, researchers occasionally came upon an unusual avatar design that preserved usability despite being nonrealistic or even bizarre.

The first example of this occurred during the process of creating an immersive city and harbor planning tool, a collaboration between Jaron Lanier and Tom Furness and others at the HITLab of the University of Washington. One of the scientists was inhabiting an avatar—a worker at the docks in Seattle—when his arm was made very large, perhaps the size of a very large crane. This most likely occurred because a designer entered extra zeroes in a scale factor in the software that measured movement. What was remarkable was that the scientist was able to pick up vehicles and other objects in a large scene at great distance with a highly distorted arm, and was able to do so with accuracy and no apparent loss of usability. This unexpected observation, and others like it, motivated an informal study of "weird avatars that were still usable."

Fast forward to the year 2014, when another happy accident led to a scientific breakthrough. This involved one of the conditions from the first experiment from above, the "extended range" condition. Recall that in this condition, when subjects moved their physical leg a little bit, they saw their virtual leg move a lot. Andrea Stevenson Won, who ran the original Homuncular Flexibility studies, had worked in pain research before coming to Stanford. After a few years in my lab, she decided to connect her previous work to VR, and dedicate her career to using VR to reduce pain. Hence she found a novel way to apply what we learned in our theoretical study with arm- and leg-swapping to help kids, specifically pediatric patients, deal with unilateral lower limb complex regional pain syndrome (CRPS).

Her work is preliminary, with a pilot study examining only four patients published in a well-established journal called *Pain Medicine*. But even though it is in the early stages, the results are encouraging.

CRPS is brutal. It's a neurological disorder caused when the central and peripheral nervous systems are not interacting properly, and it can lead to patients experiencing great pain in a particular area, for example the leg or arm. The best way to relieve this type of pain is by doing physical therapy, which is extremely painful. In our study, all four patients were receiving a variety of therapies, including physical and occupational therapy, psychological support, and medical visits. All therapies, including medications, were kept constant throughout the study period.

We added an additional VR treatment, where we manipulated the flexibility of the relationship between tracked and rendered motion in two ways. First, we altered how much effort must be exerted to take action in the virtual world by increasing the gain between participants' movements in real life and their avatars' movements in the virtual world. The idea here was to give the patients a "can do" visualization, to let them see the gain in mobility that they one day could have. Psychologists call this Self-Efficacy—the idea that one *can* achieve a goal is essential in being enabled to actually achieve it. It's clear from the medical literature that positive visualization can be a powerful aid to healing. The problem is, it's easy to tell someone to do it, but when a patient is in agonizing pain, visualizing good form in leg movement can be hard to do cognitively.

The second thing we did in our VR therapy was to alter which of the patients' limbs controlled the limbs of the virtual body, switching arms with legs. The point of this treatment was to encourage movement. In general, people prefer to pop balloons with their hands. If their arm and leg movements are switched, children must use their physical legs to move their virtual arms, and by leveraging the preference to use one's avatar's arms in this task, we can create motivation to increase physical therapy. Basically, for both treatment types, we took the academic manipulations we developed for the study we described above, and adapted them to a clinical setting.

Patients were seated and wore an immersive virtual reality headset. A sequence of balloons was programmed to appear randomly in the center of the room in front of the participant in a four-foot-wide plane scaled to the upper limits of the avatar's arm's reach. If a participant hit a balloon, audio feedback was provided in the form of a loud "pop" and the floor vibrated slightly. If a balloon was not popped within five seconds, it would disappear silently. Each patient returned to the lab for six separate therapy sessions over a few months.

The first pilot study consisted of two conditions: the normal condition, in which participants' tracked legs controlled their avatars' legs in a one-to-one relationship; and the extended condition, in which the gain of participants' leg movements was increased by a factor of 1.5, such that a moderate kick in the physical world gave the avatars' legs great range. Participant A was a 17-year-old male with CRPS in the left leg, and Participant B was a 13-year-old female with CRPS in the right leg, both right-leg dominant.

In the second study, we added a third, switched condition, in which participants' physical legs controlled their avatar's arms. Thus, a kick near waist height in the physical world would allow the avatar's arm to be raised over the head. Participant C was a 14-year-old male, left-leg dominant, and Participant D was a 16-year-old female, strongly right-leg dominant. Both were diagnosed with CRPS in the right leg.

From a clinical viewpoint, participants were remarkably calm and engaged while in the study. This was quite different from their behavior during standard physical therapy sessions, where even very slight movement was often associated with wincing, verbal statements such as "aah" and "ouch," drawing back of the limb, and pausing. In contrast, during VR treatment, they did not complain of pain, arrived eager to engage, and tolerated the therapy well, actively moving the affected extremity during the entire five-minute session in all but one case. This is felt to be significant as, during their routine physical therapy sessions, although the therapy session lasted for 30–60 minutes, the subjects would only typically be active for 2–3 minutes at a time before wincing, complaining of pain or needing to rest during an individual exercise. Subjects completed 96% of the requested activities. Some pointed out program suggestions and provided ideas for more realistic scenarios. Participants' qualitative responses were generally positive, describing the game as "cool," "simple but interesting," or "motivating, trying to beat my score from last time."[18]

Although the study described does not have enough participants to draw conclusions about pain efficacy, the measures described may provide a path for future treatment for pediatric pain participants using the flexibility provided by VR. We confirmed that VR was safe and well tolerated, and did not worsen pain or physical functioning. We also found that participants would tolerate disconnects between tracked and rendered movement. At Cornell, Andrea is spending the bulk of her days extending and testing this research paradigm. Building on our work at Stanford, she designed the portable system being used by Kim Bullock's patients. Clinicians could someday prescribe such a system for patients' at-home use.

Implementation

In spite of decades of positive clinical results, it will still take time before virtual reality is available as a medically sanctioned therapy for physical and psychological pain. It is one thing for a few pioneering researchers like Hunter Hoffman and Skip Rizzo to employ VR in the controlled settings of university hospitals and labs, and quite another for virtual reality therapies to be prescribed for use in hospitals and homes around the country. Now that the cost of hardware has plummeted, VR is in a position to be widely available to consumers; but, for its benefits to be available to the tens of millions of people suffering from chronic pain, FDA approval and coverage under insurance plans is required. More study will be necessary to designate VR as "safe and effective," the standard that will recognize VR as an approved medical device (a category that includes everything from fMRI machines and defibrillators, to tongue depressors and glasses) by the FDA. The FDA can be slow to adapt to new technology, and often looks to the existence of analogous devices when deciding on approving new technology. As a novel form of treatment, VR will need to amass a considerable amount of clinical data to meet the FDA's standard of proof.

Once a VR therapy passes this hurdle it will then need to be recognized by insurance companies, a path that usually goes through the Centers for Medicaid and Medicare Services, the $1 trillion government agency that is responsible for the insurance coverage of over 100 million Americans. The CMS requires a treatment to be "medically reasonable and appropriate" before it will offer coverage, and this will require further demonstrations of VR's clinical effectiveness.

VR has a long road before it is as much a fixture in hospitals as the TV set hanging from the ceiling. But the momentum is growing, and simply put, it is too good a use case to ignore. VR will not eliminate pain, but it is a tool, one of many, that will help.

CHAPTER 7

BRINGING SOCIAL BACK TO THE NETWORK

So far I've stressed the reality-bending properties of virtual reality, discussing how the technology can allow us—unbound by the laws of the real world—to do impossible things in virtual settings. VR's ability to create these experiences and make them feel real is one of the exhilarating things about it, and this extra-real quality has informed the most popular applications and games that have emerged in the first generation of consumer VR content. But focusing on all the spectacular solo experiences that VR makes possible obscures what I believe is the truly groundbreaking promise of the technology. It's right there in the first literary journey through virtual reality, in William Gibson's 1984 cyberpunk thriller *Neuromancer*. (Jaron Lanier coined the term "virtual reality" in 1978.) Gibson's virtual reality—what he calls "cyberspace" and "the matrix"—is defined in the novel as "a *consensual* hallucination" (my emphasis). What Gibson suggests is that it won't be the graphics or photorealistic avatars that will make these virtual worlds feel real—it will be the community of people interacting within them, bringing the world alive through their mutual acknowledgment of its reality.

People often ask me what the "killer app" of VR will be. What is going to drive mass adoption of this expensive and admittedly awkward technological apparatus? I tell them it's not going to be trips to space, or courtside seats at sporting events, VR films, cool videogames, or underwater whale-watching. Or at least, it's not going to be those things if you have to do them alone, or if, when you do share them with other people, you are excessively constrained in your ability to interact. Virtual reality is going to become a must-have technology when you can simply talk and interact with other people in a virtual space in a way that feels utterly, unspectacularly normal.

Of course, I'm not the first or only person to see the vital importance of social VR. It's no accident, after all, that the company that started this virtual consumer revolution with a substantial investment in VR was a colossal digital social network. Ever since Facebook's purchase of Oculus in 2014, creating hardware and software that can reliably achieve a high level of "social presence" for all users of VR has been the technology's holy grail, and many dozens of companies, small and large, are attempting to solve aspects of this fiendishly difficult task. And it remains one of the most difficult problems in VR—how can you put two or more users in a virtual space and allow them to interact in a human way, with each other and with their virtual environment? How do you capture and convey the subtleties of human social interaction, in the movements of the face, in body language, in eye gaze? Once again, the challenges posed by VR remind us of the richness and complexity of our human experience, because to understand how to make our avatars feel real, we have to know what we humans are doing—consciously and unconsciously—that makes our daily encounters in real life feel real. And that, philosophers and psychologists will agree, is a complicated question.

Before I get into what's so tough about creating good social VR, I want to briefly discuss why it's crucially important that those of us working on this problem get it right, and quickly. A few chapters ago, I wrote about some of the research that shows how VR can help educate and raise consciousness about the damage we are doing to the environment. But these results all rest on how VR can more powerfully deliver content that may change the way we think about the environment. What I didn't get into was how the creation of effective social virtual reality could fundamentally alter the structure of our lives and the way we communicate, namely by allowing people separated by great distances to socialize, collaborate, and conduct business in new ways.

This could have hugely beneficial effects on our quality of life, not least by reducing long daily commutes and allowing people more flexibility to work out of their office, thus freeing up time for more productive work and play. This could have great consequences for the health of the planet. If you look at the main human-caused drivers of climate change, the burning of fossil fuels is near the top of the list: nearly one-third of all CO_2 emissions in the United States is due to the burning of fuel for transportation. Reducing the amount of time we spend in cars and in planes is crucial if we want to develop a sustainable culture in the future.

Consider Shaun Bagai, a businessman in his early forties, whom I contacted after reading an article about him that appeared on the front page of the *San Jose Mercury News*. Bagai racks up about 300,000 airline miles a year as he flies around the world selling cutting-edge medical technology to far-flung clients. Over his lifetime, it has added up to over 2.5 million miles on United Airlines alone. One year, he visited 17 countries, more than 100 cities, and spent just under two-thirds of his nights sleeping in hotels. In the past decade, if you add up just his miles on United, Shaun has flown around the Earth close to 80 times.[1] Even for pampered businessmen like Bagai, travel like this is brutal.[2]

But ignore for a moment the damage all that burning gasoline and jet fuel is doing to the environment, and just think about safety. Worldwide, 1.3 million people a year die in car accidents, with an additional 20–50 million injured or disabled. In the United States, the number of deaths in 2017 was over 40,000.[3] Many of us would say that 9/11 was the worst national event in our lifetime, with nearly 3,000 people killed in the horrible terrorist attack. In 2011, cars in the United States killed more than 10 times as many as Al Qaeda did.

Then there's the daily commute, and the aggravation of sitting in slow traffic on road infrastructure that is groaning under the burden of more and more cars. In most parts of the country, investment and expansion of our highways has not kept up with population growth, adding progressively more time and aggravation on the way to and from work. This makes workers miserable and kills productivity. It may also be increasing incidents of road rage, which have grown more frequent and more deadly in recent years.[4]

Physical travel also spreads disease. When I sat in the small cabin of a cruise ship, crawling up the Alaskan shore with almost two thousand passengers and hundreds of crew members. I felt enormously lucky to have a chance to see the gorgeous coastline from the ocean, but the calming effect of the pristine wilderness was undercut by the barely contained hysteria, shared by nearly everyone on board, about germs. It was clear that the cruise line was legally obligated to encourage this fear. During check-in, while we were packed in a line of hundreds of weary travelers, stewards were handing out "cover your ass" legal notices printed on flyers alerting us to the possibility of catching horrible stomach viruses on board. On every corner of the boat, there were hand sanitizer machines. In the entrance to the dining room, ship employees dressed in bright orange uniforms politely but forcefully stepped in front of us as we entered and exited, dangling a bottle of hand sanitizer in a manner that we could not refuse. The fact is, these steel boxes and tubes we pack into as we are transported all over the planet are breeding grounds for disease. Anyone who has sat next to a sneezing, drippy, flu-ridden traveler on a plane knows what I'm talking about. Short of death and disease, travel is also quite taxing on the body and mind. Cramped plane seats, disconcerting time changes, crowded airports, unpredictable delays, and uncomfortable hotel beds are the norm for world travelers.

All of this isn't to say that VR is going to replace all travel—nothing can beat the experience of visiting a national park or a foreign city, or seeing a loved one in person. Some business deals need to be closed face-to-face. But every day, thousands of business travelers hop on planes to fly all over the country, sometimes for only short meetings. Many immediately turn around and head back to the airport so they can make it home for dinner. Do all these meetings require one's physical presence? Do the hours spent getting to airports, waiting in airports, and flying between airports make sense?

On this the climate science is clear: if we truly care about living sustainably on the planet, we're going to have to be more selective about when and where we travel. We have to realize that these experiences come at a cost, particularly as we are set to add 4 billion more people to the Earth's population by the end of the century.[5] If we truly want to find a shared, sustainable way for 11 billion people to coexist on a healthy planet, then we're going to have to figure out ways to reduce travel. Right now there is a ridiculous amount of inefficiency and waste, and it's killing our planet (and us).

No doubt this isn't the first time you've heard promises about how technology is going to eliminate travel, or allow you to work from home. Older readers might remember a series of AT&T commercials from 1993 that showed a variety of then-futuristic technologies that would soon change the world (all made possible by AT&T, of course). A few of these predictions were remarkably prescient—in the commercials a succession of brief clips showed exciting innovations like electronic books, cashless toll booths, tablet computers, and online ticket purchasing. But, for me, the real utopian moment happens in a beguiling scene showing a tanned and relaxed businessman teleconferencing on a laptop from a beach hut. "Have you ever," the commercial asks, "attended a meeting in your bare feet? You will." And indeed, we can. That technology has arrived. But the culture of the workplace hasn't been as quick to embrace the remote office.

While it's true that conferencing over phones or video applications has become more common in businesses, the stubborn fact remains that most managers still want bodies at meetings. Sales reps are still expected to travel from city to city to meet with customers. Consultants are still required to visit corporate headquarters and meet their clients in person. Academics still need to travel to conferences to present papers. Computer-mediated communications are considered okay for certain uses, but when it gets down to brass tacks, the most important encounters still need to be done face-to-face, often over a meal or drinks.

"Physical contact is important," our jet-setting businessman, Shaun Bagai, told me, when I asked him why he prefers to fly to Asia for one-day business trips when he could be using communications technology. "You can't put your hand on someone's shoulder, look them in the eyes and close a deal over a telephone or even a video conference." He went on to tell me that even when he flies to Japan, where the norm is to bow as opposed to shake hands, he goes out of his way to shake the hands of VIPs. Physical contact is a major part of how he communicates, and his job depends on successfully delivering a message of trustworthiness and competence. Business executives like Bagai will take a lot of convincing before they abandon face-to-face meetings for virtual reality telepresence. He may be able to convey his impressive medical knowledge, keen ability to present scientific data, and forward-thinking vision over a monitor, but charisma and trustworthiness are still best shared in person, over a dinner or drink.

If virtual travel and telepresence is going to replace physical travel, it will have to devise a system that allows a virtual version of Shaun Bagai to be just as warm and charismatic as his physical self. To do that the people who are creating avatars and virtual worlds, and are designing the systems that will allow users to interact with each other, will need to understand and to some degree replicate in virtual bodies the complex choreography of body language, eye movement, facial expressions, hand gestures, and physical touch that occur—often unconsciously—in real life social interactions. "Perhaps the most important problem yet to be solved," wrote Michael Abrash, the chief scientist of Facebook's Oculus division, "is figuring out how to represent real people convincingly in VR, in all their uniqueness. . . . In the long run, once virtual humans are as individually quirky and recognizable as real humans, VR will be the most social experience ever, allowing people anywhere on the planet to share virtually any imaginable experience."[6]

To get a sense of how difficult it is to capture those subtleties of human interaction that can make social VR work, let's turn to the research of William Condon and Adam Kendon. Condon was a psychology professor working at the Western Psychiatric Institute and Clinic in Pittsburgh, where, in the 1960s, he pioneered the analysis of human interaction through an innovative study of verbal and nonverbal behavior—what he called "interactional synchrony." In his research, Condon and his colleagues would film two people talking by using a special camera that could capture footage at 48 frames per second—twice that of a typical camera at the time. He and his colleagues would then exhaustively process the film by marking events as they unfolded in these fractional units of time, a system that was "similar to that used by football coaches to study films of games."[7] On each frame, the smallest details of the interaction—what was said or how subjects moved—was marked down for later analysis. Condon would note subtle movements of, say, the left pinky, the right shoulder, the lower lip, or the brows of each subject, and see how these movements were coordinated with the rhythm and content of the conversation. The point of this massive undertaking was to create a system to understand how social interaction works. His research uncovered a complex relationship between human gesture and speech. In doing so, he created a tool that others could use to study verbal and nonverbal communication, which it turned out was more complex than scholars at the time had believed. "Thus," he wrote, "the body of the speaker dances in time with his speech. Further, the body of the listener dances in rhythm with that of the speaker."

A few years later in 1970, the English psychologist Adam Kendon, working at the Bronx State Hospital at the time, used Condon's method to uncover a principle that continues to guide much of the research and the applications involving nonverbal behavior today. What Kendon discovered was that body movements—subtle shifts in posture, eye-gaze and gestures—not only dance to the rhythms of speech, but they also perform in reaction to the movements of others. In fact, these movements, both subtle and obvious, are correlated between people at a highly intricate level. In one study, Kendon videotaped a gathering in the lounge of a London hotel. Chairs were arranged in a circle, similar to how they might be in a pub, and in them sat eight men and one woman—customers at the hotel—who ranged between 30 and 50 years old. The group was encouraged to do what one normally did in pubs in the late 1960s—converse, drink, and smoke. A moderator stepped in on occasion, but for the most part the nine individuals socialized, sometimes as a part of one large conversation, other times forming smaller, sidebar groups.[8]

Analyzing the film later, Kendon meticulously mapped out—16 times per second using Condon's Interactional Synchrony tool—the movements of head, trunk, arms, and hands for all nine speakers, and created a flowchart of group movement and synchrony. By correlating all these data, he discovered that the timing of movement in social interaction is spectacularly precise. A group interaction is as choreographed—in an automatic way—as the most practiced and finessed ballet. A slight change in posture from one person causes a tiny head shift from another. A flex of the elbow denotes a change in who speaks. The choreography works within a person—with precise timing of body movements that amplify speech—but more importantly between people. The words and gestures of one person create waves in a group, controlling their gestures as well. All within a fraction of a second, and all mostly happening at a subconscious level.

Kendon is credited for discovering how complicated the coordination of group interaction is, and how precisely choreographed nonverbal and verbal behavior is between people. Interactional synchrony, he showed, is the secret sauce of social interaction. But it can vary across conversations—sometimes synchrony flows, and in other conversations it is lacking. The amount of synchrony can reflect the quality of a conversation. To quote Kendon, "To *move* with another is to show that one is 'with' him in one's attentions and expectancies. Coordination of movement in interaction may thus be of great importance since it provides one of the ways in which two people signal that they are 'open' to one another."[9] Synchronous conversations work better than ones that lack nonverbal rapport. On the other hand, when technology interferes with synchrony, the conversation suffers. A 20-year-old study from 1996 on videoconferencing conducted by a team of psychologists from London examined the effect of latency. They had 24 pairs of subjects perform a task over videoconference. Half of them had high latency—about half a second. The other half had very low latency. The pairs had to perform a task together using maps. Two results emerged—first, participants in the delay condition made more mistakes in the task. The lag in conversation actually made productivity suffer. Second, the speakers interrupted one another more often during high latency. Most of us have had this experience during bad cell phone or Skype conversations. Delay harms the "flow" of the conversation, or in Kendon's words, the synchrony.[10]

Since Kendon's landmark work in the late '60s, dozens of studies have examined the effect of interactional synchrony on outcomes. A study from the late 1970s looked at about a dozen college classrooms over time, and demonstrated that students and teachers with high nonverbal synchrony in posture had better relationships than those with low synchrony. They rated themselves as more compatible, more together, and higher in rapport. [11]

Today, we no longer have to laboriously analyze film stills to study synchrony. With VR, it's possible to get much more accurate measures of body movements. Tracking systems like Microsoft's Kinect can detect and record subtle body movements in three dimensions, collecting data that can then be analyzed by computers. In one study, we used the Kinect to capture the micro-movements of many pairs of people as they collaborated on a brainstorming task. Because the process was done via computer as opposed to the more painstaking methods employed by Condon and Kendon, we could look at a much larger sample—50 groups instead of just a few— and examine data from interactions that lasted longer than those in previous studies. We then correlated those micro-movements between subjects to produce a "synchrony score." People whose body language displayed more synchrony received a higher score than those with less synchrony. When we compared the synchrony scores of pairs against each other and then compared pairs' performances on the brainstorming task, synchrony predicted rapport again. In our studies, high synchrony pairs came up with more creative solutions to their task than those with low synchrony. They actually performed better. Nonverbal synchrony is a marker of a good, productive conversation.[12]

But the advantages of synchrony don't need to be passive. Synchrony can be implemented by design.

Why are soldiers forced to march together? Why do people in churches sing and clap in unison? It turns out that not only is nonverbal synchrony an outcome of good rapport, it can actually cause positive outcomes. In 2009, researchers at Stanford manipulated synchrony by asking subjects to either walk in step with one another or alternatively just walk normally. The study showed that people who moved synchronously demonstrated more cooperation and were more generous to one another than those who walked normally. Dozens of studies have demonstrated this effect—nonverbal coordination contributes to greater social cohesion and group productivity.[13]

Philip Rosedale likes to tell a story about how he got into creating virtual worlds. "When I was really young I was just obsessed with taking things apart and building things—with wood or electronics or metal or anything else."[14] When he was a teenager he decided he wanted his bedroom door to open up automatically, like the doors he saw in Star Trek. So, much to his parents' displeasure, he cut through the joists above his ceiling and installed a garage door opener in the attic so that this whimsical teenage dream could come true. Later, in his twenties, when he started thinking about what he wanted to do with computers and the brand-new Internet, he would think back on his love of tinkering, and he built one of the most innovative and disruptive platforms in the history of the Internet.

How many of us, he wondered, have amazing or crazy ideas about things we'd like to do, but we don't pursue them because we don't have the resources, or the time, or the knowledge to make them happen? Perhaps, he thought, the Internet could be a place where people could make those dreams happen. More importantly, they could do these things together, cooperatively.

Over the past two decades, Rosedale has dedicated himself to building virtual worlds. He began with the popular online avatar playground, Second Life, which launched in 2003 and became the most popular—and written about—online world of its kind. Second Life realized many aspects of Philip's vision for a metaverse—its servers allowed large groups of users to socialize together online, and in their world they could build, or buy on Second Life's marketplace, everything from land and houses to jewelry and clothes for their highly customizable avatars. But all this customization and possibility came at a price: the dense menus and complicated keyboard- and mouse-driven interface was a big hurdle for a vast majority of the over 43 million users who registered accounts with Second Life during its first 12 years.[15] This, Rosedale believed, is one of the main reasons that Second Life didn't continue to grow. The number of regular users is around one million—a healthy amount of people, to be sure, but not the sprawling and crowded metaverse that Rosedale imagined.

Rosedale's energies are almost maniacally focused on developing sophisticated gestures and social cues for his avatars. I got a glimpse of this when I first visited his High Fidelity offices shortly after they launched, when Rosedale's long-anticipated new project was first unveiled. The event had taken over what looked like a typical startup space—bicycles in the lobby, high ceilings, whiteboards and laptops everywhere. The only noticeable difference was an unusually high number of Oculus developer kits, and infrared tracking cameras located at each workstation.

After grabbing a drink from a makeshift bar—Mojitos were the specialty—I began wandering around the large open room. About a dozen employees wearing costume-like lab coats were also moving about, demonstrating the newest features of the virtual world to the guests. At first glance, High Fidelity looked like Second Life with better graphics. But it immediately became clear that there were crucial differences. First, there was the fact that while High Fidelity, like a traditional computer game or virtual world, could still be viewed on a monitor and navigated with a controller, or mouse and keyboard, it was clearly being designed with VR in mind. Users could inhabit the bodies of their avatars (rather than observe them over the shoulder, the default mode in Second Life). And like Second Life, it could support live audio, allowing you to chat with other users. But now these sounds were spatialized, rendered in three dimensions.

But the truly groundbreaking difference was High Fidelity's

intuitive interface, which removed clumsy mouse and keyboard controls, and allowed their avatars to reflect their real-world hand gestures and facial expressions. Avatars in High Fidelity's virtual environments, for example, had realistic fingers that users could control by wiggling them in front of a sensor mounted on their HMD. The facial expressions and gestures of realistic avatars were captured by a depth camera that sat on top of the monitor and mapped the body movements of players to their digital doppelgangers. I put the HMD on to see what it was like to step into High Fidelity, rather than observe it through the framed screen of a computer monitor. The scene created by all the hardware was a basic cityscape in which avatars could walk, fly, gesture, and talk to one another. Moments after logging in, I forgot about the people milling about me, sipping their cocktails in the physical room— the virtual people were the ones I was looking at, gesturing to, and listening to. The transformation was instant and thorough.

As with Second Life, Rosedale's ambition for the size of High Fidelity is almost boundless. Rosedale makes clear he wants it to be nothing less than the metaverse that has been predicted in the sci-fi novels of William Gibson and Neal Stephenson. "Virtual reality is the next disruptor for society, after the smartphone and after the Internet," Rosedale has said. "Much of our human creativity may move into these spaces. I think that it will. We will move into the metaverse for much of our work, design, education, and play in the same way we moved to the Internet."[16] He imagines a virtual world that could exceed the size of Earth. Where people in Alaska can visit a virtual New York City and spend a night on the town, and where people in the real New York City can climb Denali to get away from the pressures of city life. To achieve the scope of this goal, he wants to forgo the use of centralized servers—the standard procedure for hosting online worlds—and outsource the computing power needed to run his metaverse onto a growing network of users' home computers and mobile devices. "We'll have 1,000 times more machines available," he said in an interview in 2015. "If you could use that, how big a metaverse would that be? You wind up with the rich details of a videogame. You get immense spaces with local details. If we use all these computers we have today, the virtual space could be the size of the land mass of Earth today." This would save millions of dollars for the new company, and have vast implications for the energy consumption model.[17]

One thing I noticed when I first tried High Fidelity was how often I looked down at my hands, and how novel it was to see them moving as I interacted with other avatars. We have hand tracking at the lab, of course, but in terms of manual expressiveness you might as well be wearing one of those "We're #1" foam fingers you see at sporting events—most avatars, even with advanced motion controllers, are still big clumsy appendages that barely approximate the possibilities enabled by fully articulated human digits. Rosedale is obsessed with hands. He sees them as the key to the intuitive interaction that will drive people to experience virtual worlds. "Using your hands to manipulate things," he likes to say, "is the core part of human experience." And hands aren't just used to manipulate objects, they are also our primary way of initiating physical contact with other people. This leads to one of Rosedale's favorite questions: "What is the virtual handshake?" This question is meant to promote a discussion about the sizable gap that still exists between a compelling real-world encounter and the kind of social experience that exists in virtual reality. The virtual handshake is Rosedale's metaphor for effective social presence, the kind of experience that can save the Shaun Bagais of the world some of those exhausting trips, and truly draw people into virtual social worlds.

THE VIRTUAL HANDSHAKE

This notion of the handshake, and its virtual future, is one I think about often. I've been working on some aspects of this social VR question for years, and in June 2010 I visited Philips Corporation, based in Holland. Philips is one of the largest electronics companies, employing over 100,000 people in dozens of countries around the world; their technology has been a fixture around the globe for more than a century. Philips funded research in my lab to produce virtual encounters that were as compelling as—or even more compelling than—face-to-face ones.

Indeed, for my collaboration with Philips, flying from Stanford in California to Eindhoven in Holland demonstrated all of those features of travel—the flight time, with a stopover, is three-quarters of a day, the time change is nine hours, and the lower back gets punished. (And yes, all those hours in the sky gave me plenty of time to ponder the irony of repeatedly flying to Holland in order to develop a system to reduce physical travel.)

The goal of the project with Philips was to explore ways to convey in virtual encounters something approaching the intimacy we find when we speak to someone face-to-face. We were looking at many technologies and testing their psychological effects. Current VR technology, of course, can't transmit all the subtle cues that are conveyed when we interact with others, but it does offer some interesting alternatives. In many cases the physical cues we see when we speak with someone are reflections of involuntary physiological reactions—a blush, a nervous tic, or a genuinely happy grin. These types of responses are often accompanied by changes in a person's physiology. What if we conveyed the underlying physiological responses of a person into the virtual environment?

For example, one system we built allowed people not only to see

one another's avatars but to see data visualization of their con-
versational partners' heartbeats. Imagine asking someone out on
a first date. You are 80% sure your proposal will elicit a resound-
ing yes—but that 20% chance of rejection is not negligible. As
you (and your avatar) make your pitch, you get to see everything
another person might see via Skype or a high-end videoconferencing system—your partner's facial expressions, body posture, and
voice. But you get an extra layer of information—a biometric read-
out of her heartbeat. You get to see her heartrate accelerate as the
words from the rehearsed date request formulate on your tongue.
In other words, the virtual interaction gives you a tool you wouldn't
have had face-to-face, and you can craft your delivery to match her
body rhythm. Our research at Stanford looked at people's reactions
to perceiving their partner's heartbeat, and showed it heightened
intimacy. People felt "closer" to someone when they could glimpse
that person's physiology.[18]

The other part of the project with Philips was building virtual
social touch. In addition to me, Philips had reached out to another
one of the handful of scholars who had rigorously investigated the
psychological implications of sending touch over the Internet: my
colleague and friend Wijnand (pronounced Vi-Nan) IJsselsteijn,
who is a professor at the Eindhoven University of Technology.
Wijnand is an expert at building haptic devices that can measure
touch on one end using sensors that detect body movement, and
ones that can create the sensation on the other end using electric
current, vibration, motors, and air puffs. He is also an expert at testing to see if they are effective at eliciting psychological intimacy.

Touch can be a powerful dimension of social life. Studies show
that waitresses who touch customers on the shoulder can influence
them to order more drinks and also get bigger tips than ones who
don't. (But only if they're not being flirtatious and definitely not if

it inspires jealousy from another customer at the table.) This so-
called "Midas Touch" effect is well documented in psychological
literature that goes back to the 1970s. Wijnand and I wondered if
virtual touch worked the same way. This effect is a particular interest of Wijnand's, who has worked for almost a decade to answer
this question. He has spent so much time working in this area, in
fact, that he was inspired to name his youngest son Midas.

In the first study to look at the Midas Touch in virtual space,
Wijnand had two people chat online using instant messaging. One
was the subject, while the other was a "confederate," a person who
is part of the experiment but pretends to be another participant
in the study. During the interaction, experimental subjects wore
a sleeve on their arm that used six points of vibration contact to
simulate being tapped on the arm. Some subjects received the
virtual touch from the confederate, who could activate the vibrating sleeve on a networked computer, and others did not. After the
experiment was over, the confederate got up from the computer
station in front of the subject and dropped eighteen coins on the
ground. Wijnand measured whether or not subjects helped pick
up the coins, with the prediction that those who had been touched
virtually would be more helpful than those who had not. He demonstrated that a majority of those who had been touched helped,
while those who did not get touched helped less than 50% of the
time. In other words, the Midas Touch worked virtually.[19]

Wijnand's work is important replication work. In other words,
it shows that virtual touch can bring about the same benefits that
physical touch can bring, by replicating a finding that is fairly
accepted by scholars worldwide. The work conducted in my lab
took a different approach by examining some benefits of touch that
can only occur in the virtual world. We looked at mimicry. It's long
been known that nonverbal mimicry causes influence—simply by

subtly matching another person's gestures you can get him to like you. Tanya Chartrand, while a professor at New York University, was perhaps the first to provide rigorous data on this "chameleon effect." She had people go into job interviews and mimic the nonverbal gestures, such as leg crossing, of the interviewers. People who mimicked the interviewer were more likely to get favorable evaluations than those who did not, even though the interviewer had no conscious idea he was being mimicked.[20]

Have you ever shaken your own hand? Mimicry via touch might also provide influence. To test this, we bought a rubber hand from a Halloween costume store and attached it to the stick of a "force feedback joystick" that uses motors to play back movements that are recorded by another similar joystick. We actually built Rosedale's hypothetical conversation piece—a virtual handshake—by having one person hold the rubber hand and move the joystick, and having another person feel the movement on a similar handshake machine. Two people met face-to-face in the lab but never touched physically. We then had them virtually shake hands using our newly crafted machine. But one person never actually felt the other person's shake. Instead, he received his own handshake that we had previously recorded, and was told it was the other person's. In other words, he shook his own hand.

People who received their own shake liked their partners better—they treated those "digital chameleons" more softly in a negotiation task and also rated them as more likable—than those who received the other person's actual movements. People had no idea they had been mimicked, but the subtle effect of familiar touch was a winner.

Imagine giving a talk to a thousand people, trying to convince them of some ridiculous notion—say, for example, why they should travel virtually instead of physically. I do this from time to time.

Perhaps, if I could deliver a personal Midas Touch to all the people in the audience, they would be more receptive to my message. Unfortunately, it would take me hours to shake every single person's hand physically. But if each person had the application that Philips Electronics is devising, which uses the accelerometer and vibration motors in today's smartphone to simulate touch, then I could shake all of their hands at once, leveraging Wijnand's experimental findings. But, even better than just tapping them on the shoulder, I could send different versions of my handshakes to all of them simultaneously, scaling up my ability to mimic. It would be a politician's dream.

The human face is made up of over forty muscles and is capable of an astonishing range of expressions. Until recently the palate of emotions available to avatars was quite small—not much more than you'd get in a list of emoticons on your phone messaging app. But in 2015, a company called Faceshift, made up of a small group of Swiss scientists, seemingly cracked the code on face tracking and rendering, and did so through an ingenious method.

They developed a real-time system that could "track" facial movements. About 60 times per second, an infrared camera would scan the face. Based on the time it takes for the light to bounce from the camera to each point on the face and back to the camera—the time of flight—it is possible to create a depth map of the face. That part isn't new; the Microsoft Kinect had been doing it for years. What was new about Faceshift was their software's ability to read gesture and emotion. They decided to categorize the face into 51 categories, called morph states. For example, in what percentage is the left eye open? How large is a smile? When a certain set of movements formed an expression, these changes could be measured and categorized as a particular emotion. So every time the infrared camera scanned the face, it returned a value on all 51 morph states. Reading this book right now, you likely have low values on most states. Perhaps a bump in eyebrow movement, as you furrow your brows in concentration. But compared to laughing and talking with friends, there is likely not too much facial movement as you read.

The next step in the equation is to render the morph states onto an avatar. The Faceshift team designed their avatars to quickly and effectively show animations along those morph states. Previous techniques I had seen over the years took a different approach, more based on a bottom-up rendering of micro-movements as opposed to a top-down categorization of emotional states. The difference was jaw-dropping.

I'll never forget the first time I experienced this technology in action. A colleague was sitting at a desk, looking at a computer. On the computer was a highly photorealistic 3D model of his face that had been scanned in by the program in a matter of minutes. It looked just like him, and was as real as any picture or video I had seen. But more importantly, it moved like him, in real time. When he raised one eyebrow, so did his avatar. When he laughed, the avatar performed *his* laugh—not just any generic laugh, but a laugh I would have recognized anywhere. Even if the avatar face hadn't looked just like him, I could have picked him out of a lineup in no time at all, just based on his distinct gestures.

As he and I talked about the technology, a strange thing happened. I stopped addressing him. Slowly, over our conversation, my body pose had oriented away from his physical body and toward his avatar on the screen. I was being completely rude to my colleague—not looking at him and violating the unspoken rules of personal space as I turned my back to him. On the other hand, I was being very polite to his avatar. At some point in the conversation we realized what was happening. We took a picture, and sent off a congratulatory e-mail to some other colleagues. Avatars had just become real.

Over the next few months, I was able to watch this technology migrate from a desktop computer to a tablet and smartphone. Imagine using your phone to Facetime or Skype a friend, but instead of seeing a video feed, you see your friend's avatar face doing just what her physical face is doing, in real time. At first I had mild expectations for this "phone avatar" system, but my mind was utterly blown by how social it was. Apple Computers Incorporated must have been similarly blown away; they bought the company Faceshift in 2015.

One reason we might prefer avatars to video for communication is latency. Video cameras are not very discriminating—they duly record everything that's going on and do not differentiate between what is important and what isn't. When you are communicating over traditional videoconferences, for instance, every single pixel, over every single frame, is transmitted over the network. Think about that—the lamp behind the desk behind you travels over the network every single time the screen updates. Even with good compression algorithms it is horribly inefficient. In a conversation, we should pay attention to the gestures of the other person—where are they looking, are they smiling, is there a telling twitch of the mouth. But videoconferencing at its essence is designed to send everything the camera sees over the network, regardless of how important the feature is concerning communication.

The neat thing about VR is that you don't need to send all those pixels over the network over and over again. The way that Second Life worked, and the way High Fidelity and the other newer versions of social VR will work, is that all of the 3D models are stored locally on each person's machine. Look around wherever you are reading this book. Your room might have a chair and a table, or maybe you are on a train. In VR, all those objects—including you—are captured as a 3D model. Those models—or at least the ones you choose the other person to see—are stored locally on the computers of everyone who is in the VR chat.

Virtual reality functions in cycles—the computer figures out what someone is doing, then redraws his or her avatar to show changes based on that behavior. For example, as a person in Cleveland moves his head, smiles, and points his finger, the tracking technology—let's say the Faceshift system—measures those actions. As the Cleveland person moves, the computer of his buddy in Tuscaloosa, which already has a photorealistic avatar of the Cleveland guy, receives that information over the Internet and modifies the avatar to make it move, too. Tracking the actions of two speakers, transmitting them online, and applying them to the respective avatars all occur seamlessly, and all the participants feel as if they are in the same virtual room, in a movie together. In social VR, tracking equipment detects what a person does and instructs the other person's computer to redraw the avatar performing the same action. Everyone's computer sends the other machines a stream of information summarizing the user's current state.

It turns out that the tracking data—the only thing that needs to be sent over the network—are almost negligible in terms of bandwidth compared to sending a high-resolution image. All the pixels are already stored on each computer. Consider the Faceshift example—a very dense, photorealistic model is already stored on the machine. It is roughly thousands of times the size of a string of 51 numbers—the text required to represent the morph states of the avatars. So the latency is reduced, since you are sending tiny text packets over the Internet instead of monstrous pictures. The outcome is that you get to see your friend's gesture *when* the gesture occurs, not precious fractions of a second later.

In June 2016, Tom Wheeler, then the chairman of the Federal Communications Commission, spent 90 minutes visiting the lab. He was fresh off a victory in a big court case for Net Neutrality. After a discussion of some of the privacy issues posed by VR, we turned to the subject of bandwidth. Even with the most efficient data transmission, the possibility of hundreds of millions of people spending lots of time working and socializing in VR presents challenges to our communications infrastructure. What, Wheeler wondered, would the government need to do to make a metaverse work? The best metaphor I could give him was a telecommunications version of commuter rush hour. With widespread VR use, there will periodically be massive bursts of file transfers as people send their avatars, the 3D models from their living rooms, etc., over the network. But once these relatively brief "rush hour bursts" are over, they will be followed by very light traffic as a conversation proceeds, simply the back-and-forth of the tracking data. The challenge, however, is that rush hours are predictable—you can expect a quagmire on weekday mornings and evenings. But VR usage is likely to be less structured. Of course, there are no policies or norms yet surrounding VR, but Wheeler believes he now needs to fathom a system that will have intense bursts of bandwidth followed by prolonged light traffic.

For decades, I had heard the arguments about how avatars will be more efficient than video and eventually reduce bandwidth (and latency), but I didn't truly appreciate this until I saw the Faceshift system for the first time on a phone display. Because locally saved avatars don't have to travel over the network, the images themselves can be very detailed. They won't look like cartoons—they will be super high-resolution models, with perfect lighting effects for shadows and reflections. They will dwarf the fidelity of the networked faces you see on Skype, Facetime, or other videoplatform conferences, which need to keep the realism level suppressed in order to prevent latency. The visual quality of the networked Faceshift avatar simply blew me away. It was the "realest" face I had ever seen networked.

Another unanticipated benefit became clear within minutes of using the system. I often Skype with my mom, who lives across the country, and she is typically holding a tablet while we talk. It's massively challenging for her to hold the tablet in a way that keeps her face in the center of the frame. Indeed, my kids have spent many an hour conferencing with their Nonna from the nose up, as the bottom of her face is cut off from the frame. In VR, the avatar face is always perfectly centered. Which is great for the viewer, but also liberating for the other person, who no longer has to dedicate effort to holding the phone at awkward angles to keep herself centered. This is one of those features that seems small—but in practice is night-and-day. It's really hard to go back to video after this.

And of course, avatars solve the eye contact problem that exists in video chat. When we are using a program like Skype, our eyes are naturally drawn to the face of the person we are conversing with. The problem is that the camera is not located in the middle of the screen, and so you are NOT looking at the camera, which is usually on top of the computer monitor. So the other person sees you as looking down, not right at her eyes, which only happens if you look right at the camera. This so-called eye contact problem has plagued videoconferences for decades. Many of the obvious solutions—for example, putting the camera in the center of the screen—have been tried, but to date none has really worked.

With avatars, eye contact is solved completely. The avatar is a 3D model which can be oriented in the proper direction with simple trigonometry.

But perhaps the best reason to prefer communicating through avatars over video gets down to our own vanity. Videophone technology has actually existed for more than half a century, but it has only been recently, when the Internet drove costs down to virtually zero, that it has been widely used. One would have thought that the advantages of seeing the person you are talking to would have made it a more popular means of communication earlier (or a more popular mode of communication than it is today). But it appears we like the convenience of talking with someone whenever we want, more than the inconvenience of making ourselves visually acceptable for those conversations. The novelist David Foster Wallace parodies this idea brilliantly in his novel *Infinite Jest*, which imagines a future in which videophones replace telephones only for people to realize how stressful it is to be seen and heard at the same time. Traditionally, Wallace writes, phones "allowed you to presume that the person on the other end was paying complete attention to you while also permitting you not to have to pay anything even close to attention to her . . . you could look around the room, doodle, fine-groom, peel away tiny bits of dead skin from your cuticles, compose phone-pad haiku, stir things on the stove; you could carry on a whole separate additional sign-language-and-exaggerated-facial-expression type of conversation with people right there in the room with you, all while seeming to be right there attending closely to the voice on the phone."

In Wallace's satirical world, people begin donning masks that make them look more attractive while they use videophones, and as these outfits become more idealized and unlike their real-world appearance, it results in "large numbers of phone-users suddenly reluctant to leave home and interface personally with people who, they feared, were now habituated to seeing their far-better-looking masked selves on the phone."[21]

Infinite Jest was published in 1996, when the Internet was still in its infancy, and did not anticipate a world of malleable digitized avatars, but Wallace's intuition about human self-presentation and vanity was spot on. We see this already in the carefully crafted on-screen personas that people use on social media and dating sites. And we will see it in this world of avatars as well. Even the slightest tweaks of an otherwise lifelike avatar can have noticeable effects on how you are perceived. In 2016, we studied this in relation to the smile. There are well-established findings about the positive effects of smiling on social interaction. What would happen, we wondered, if we leveraged the flexibility available to digital representations and *enhanced* a person's smile during a conversation in virtual reality?

In the study we tracked and mapped the facial expressions of a participant during real-time conversations with other participants in VR. As these conversations happened, we either enhanced the smile of the participant's avatar or rendered it truthfully. Analyses using the Linguistic Inquiry Word Count (LIWC) revealed that participants who communicated with each other via avatars that exhibited enhanced smiles used more positive words to describe their interaction experience compared to those who communicated via avatars that displayed smiling behavior reflecting the participants' actual smiles. Not only that, participants reported that they felt more positive and had more social presence after experiencing the "enhanced smile" condition than they did in the "normal smile" condition. These results are even more striking when you consider that fewer than one in ten of the participants were even able to consciously detect the smiling manipulation.[22]

Avatars do more than just represent conversationalists, they change them. Happier avatars make for happier people.

There are subtler and more insidious aspects of this kind of transformed social interaction, some of which I've written about elsewhere. To put it bluntly: we prefer those who look and sound like we do, and so experiments in which we've modified the appearance or sounds of avatars to more resemble the actual appearance of those they are addressing, have found that users perceive those avatars to be more attractive and influential than those who look different. We should expect to see a lot of manipulations of this kind in a world of communication mediated by avatars in virtual worlds. It's the logical extension of what we already do in social encounters, where we change our clothing, our speech, our body language, and other aspects of our self-presentation depending on the situation we are in. We dress and act much differently in a job interview than we do when we are hanging out with our friends in a nightclub.

We will also need to keep an eye on how people behave in virtual environments. The past few decades have seen the gradual dismantling of the utopian hopes of early Internet users that online social spaces would develop as a kind of information-rich digital agora, where ideas would be freely exchanged and intelligent conversation would flow. However, the anonymity that was supposed to protect people's speech and identity has also made possible a powerful subculture of Internet trolls who take delight in making other people miserable—sometimes because they disagree with what a person is saying and sometimes for sport. If innovators like Mark Zuckerberg and Philip Rosedale are right, and public debate shifts to virtual spaces, we can only guess as to what these encounters will be like. Avatar bodies cannot hurt you, but their perceived physicality and voice-carrying capabilities do make a greater impact than a comment on an article or a Tweet, and indeed there have already been allegations of sexual harassment and avatar trolling in some early virtual environments. The good news is that a troll could be made to go away, either by being banned from a space or by being blocked by a user. Having experienced the bruising vitriol of trolls myself, however, I am not entirely sanguine about the future.

But I hold out an optimistic hope that good social VR might actually improve things. It is much easier to dissociate a person from her basic humanity and the respect she deserves when you are perceiving her through short text messages. If we start seeing people online again as humans and are able to achieve, in some new way, the elements of synchrony that help bind us to other people, it might improve the dialogue online and open up a more productive and civil public space.

CHAPTER 8

STORIES IN THE ROUND

The water is rising. Wind roars through your ears and you can feel the rooftop underneath your feet shake. In every direction you look, the floodwaters are slowly creeping up. Through the driving rain you can see your neighbors, screaming in distress, standing atop their homes, as you are, desperately looking for help. A body floats by, a poor soul not lucky enough to have a rooftop. You hear, and then see, a helicopter fly overhead and you wave and wave but it passes by. When will the water stop? What happens if nobody comes to help you?

This harrowing experience was designed in collaboration with NPR journalist Barbara Allen, who approached me in 2012 to see how VR could be used to heighten the storytelling powers of journalism. We at the lab had been thinking for years about the potential of immersive journalism, but until Barbara knocked on my door, we didn't have the time, motivation, or journalistic expertise to create a simulation. It took us a while to find the right scenario for the project and we discussed several possibilities. Eventually, Barbara came up with the brilliant idea to simulate the aftermath of Hurricane Katrina, so that users could get a better sense of the human terror and suffering caused by the storm, experiences that traditional media could only hint at through distant camera shots or in written reports. Barbara had covered the storm, and done a lot of documentary work, so she was well acquainted with the kind of details that could bring this terrible scenario to life. Not to mention that the physical tracking space in the lab at Stanford was about the size of an actual rooftop, so the experience would really leverage the space. What followed was an iterative process: we'd pore over Barbara's notes and videos, program the visual scene and interactivity to match, and then repeat the process, often bringing in new journalists to give us feedback and then going back to the drawing board. The project culminated in a large "opening" where a number of prominent visitors were able to experience Katrina. Indeed, when James Harding, the head of BBC News, and Marty Baron, the editor of the *Washington Post*, visited to learn about VR, this was the centerpiece of their learning experiences.

At the time we were doing our Katrina experiment, we couldn't have imagined that actual examples of VR journalism would be created and widely disseminated only a few years later. But almost immediately after consumer VR became possible, reporters, news organizations, and independent producers seized the moment to begin producing original journalistic content. No organization has been more bullish than the *New York Times,* which distributed over a million cardboard viewers to its print subscribers and created a high-end, VR-specific smartphone application to distribute *Times*-created VR experiences. Other news organizations, like VICE, the *Wall Street Journal*, PBS *Frontline*, and the *Guardian* in the UK have also experimented with the medium. Bringing audiences closer to the reality of a story has always been the preoccupation of journalists, and VR, it seems, offers an ideal multimedia experience. There is a hint of urgent and perhaps forced optimism, too, as traditional journalism outlets try to figure out how they can recapture audiences they have been losing to the proliferating sources of news in our increasingly fragmented media world—why pay for it when you can get the news for free on your browser or Facebook feed? In this light, adding a new exclusive VR feature to the suite of services offered by the likes of the *New York Times* may be one way to address the financial woes that have beset the ailing news business.

After all, new media technologies have always gone hand-in-hand with journalism; the definition of journalism has constantly evolved along with those technologies. Beginning in the seventeenth century, early newspapers consisted of just the written word, but as soon as improved printing technologies allowed it, engraved illustrations and diagrams were added to written accounts. By the latter part of the nineteenth century, photographs, or detailed engravings based on them, began to appear alongside news stories. These new images represented the world in rich detail, and in doing so felt "real," promising an authenticity to contemporary audiences previously unimagined.

Of course, how "truthful" these photographs actually were is questionable. Beyond the fact that photographs only capture reality from a particular point of view, we know that early photojournalists frequently staged their pictures. Early on, taking a photograph was a time-consuming, expensive, and burdensome process, and these pioneering photographers would have wanted to get their image absolutely right before taking the shot. The Civil War photographer Mathew Brady, for instance, is known to have rearranged the bodies of fallen soldiers on battlefields to improve the composition of his pictures.[1] No doubt this sleight of hand felt justified to Brady because by increasing the visual impact of his photographs of the carnage, he felt he could better convey the horrors of these bloody battles. But such manipulation would clearly be a serious violation of the professional ethics that guide photojournalists today, when even adding a digital filter to manipulate the brightness in an image is condemned.

In the twentieth century, radio, newsreels, television, and later the Internet all continued to shape the news business with new multimedia and interactive features, each innovation raising questions about how to uphold journalistic ideals of objectivity, independence, and truth-telling that began to develop in the mid-twentieth century. Nothing has challenged these principles more than the Internet, which has irrevocably changed—for good and ill—the distribution model of news, allowing consumers more choices while at the same time undermining the authority of traditional news organizations. Increasingly, we see the very idea of objectivity brought into question by the fragmentation of the media and the diverging depictions of our reality.

Consumers of news are increasingly drawn to news sources that confirm their own beliefs. As the number of news organizations that attempt to pursue the established ideals of the profession decline, their deeply reported pieces can get drowned out by stories presented from partisan perspectives or even fake news designed to grab attention and generate outrage. Add to this situation a distrustful and cynical population that no longer can distinguish between the work of journalistic professionals and propaganda, and you have a situation ripe for mischief.

In this fast-changing news landscape we have two big reasons to be wary of a medium like virtual reality. One, it plays powerfully upon our emotions, an affordance that in many situations doesn't exactly encourage rational decision-making. A viewer who exposes himself to a VR depiction of an atrocity, for instance, will feel like a personal witness to that event and experience the resultant level of outrage. Where does that outrage go? Stoking these kinds of emotions and playing upon people's instinctive desire to lash out at perceived threats is a time-worn strategy for tyrants, terrorists, and politicians. I have little doubt that virtual reality will be an excellent tool for spreading propaganda.

A second, related worry about VR is that, due to its digital nature, it can be easily altered and manipulated. This doesn't make it different from other media, of course. We know photos and video are subject to manipulation, and even written accounts can be biased to present an ideological point of view. The fact that it is common for other media to be strategically altered for deceptive ends shouldn't give us comfort. In VR, which actually "feels real," the potential dangers for misinformation and emotional manipulation are exponentially greater. VR will force designers to make decisions simply based on the technology. What should the eye-height of the user be in the scene, her actual height or a standard height based on where the camera should be placed? When watching TV, we all are the same height, the height of the camera. Should the same be true in VR? Then there is stereoscopic view, which requires all sorts of small decisions to be made about how to match the content to the user's eyes. It is very difficult to match the simulation to create a perfect stereoscopic view. People vary in their ability to fuse a scene—some of us are even "stereo blind"— and the distance between people's eyes, the so-called interpupillary distance, varies greatly, while the hardware does not adjust all that easily. These features matter greatly to perception in VR, and there is no doubt that decisions on these parameters will change how consumers interpret news. When false events are put forth, it will be hard to argue with viewers who believe them. After all, they will have seen them with their own eyes.

Ethical journalists will respect the standards that evolve to ensure accuracy and objectivity in VR journalism, but it's safe to say some purveyors of nonfiction narratives will abuse the mutable nature of VR for ideological or sensational ends. Initially, these abuses of VR technology will occur with immersive computer-generated VR experiences, which have to be built from the ground up and are entirely within the control of the producer. Because they are computer-generated they still lack the detail and photorealism that can easily fool people into believing they are real. However, this may all soon change with advances in "light field" technology, which allows digital cameras coupled with powerful computers to extrapolate enough information from the light they receive to build photorealistic volumetric avatars and place them in three-dimensional spaces.[2] We can imagine a future when photorealistic VR video can be quickly recorded and edited after the fact with the same ease as photo editors shave pounds off models on magazine cover images. When this happens, it is not difficult to imagine unscrupulous creators of VR "news" tweaking footage to their ends, just as Soviet leaders were able to disappear discredited apparatchiks from historical photos, or as political campaigns in the United States have doctored images of campaign rallies to give the appearance of larger crowds.[3] In fact, the demo video for Lytro, one of the companies leading the development of light-field technology, wryly nods at this very kind of manipulation, showing an astronaut walking on the surface of the moon before the lights come on, the camera pans back, and we see a Stanley Kubrick lookalike sitting in the director's chair choreographing the whole scene.[4] It's a clever callback to one of the great conspiracy theories of all time—that the US government hired the director to fake the moon landing.

The pioneer of using computer-generated and volumetric VR for journalism is Nonny de la Peña, who has created a series of fully immersive experiences that allow viewers to witness real-world events like the controversial Trayvon Martin shooting and a tragic domestic homicide case that occurred in South Carolina.[5] A trained journalist who was drawn to VR for its immediacy and empathy-encouraging qualities, de la Peña does everything in her power to make these experiences as fact-based as possible, building them off of meticulous research that includes eyewitness statements, crime scene photographs, architectural drawings, and audio recordings. Very little that is built in a de la Peña scene is imagined. The Trayvon Martin story, for instance, does not depict the shooting at the scene because there is no reliable account of it. Instead, de la Peña takes us into the homes of nearby witnesses and allows us to stand in the room with their occupants at the moments that they heard the gunshots and made calls to 911.

De la Peña knows she has to go to great pains to assuage the concerns of critics who worry about the issues raised above, and when I spoke with her she was engaged in a project with the Knight Foundation to define best practices for VR journalism. But she is also quick to point out how recently documentary films were at the receiving end of these concerns. All media, she has pointed out, are artificial, and engage in rhetorical tricks and techniques to heighten engagement with the reader and viewer. As she told an interviewer, "When you're making documentary films, you have to cut to people working in the field, or cows, or reflections on cars. You're just not really showing anything that's happening based on what people are saying. Somehow, people feel that doing that in a VR way can be less ethical. But it isn't. It's just a different approach."[6] It was only a few decades ago, she points out in the same interview, that this debate raged over Errol Morris's creative use of filmed re-creations of a 1976 police murder in Texas in his Oscar-winning documentary *The Thin Blue Line.* Today, we see

such filmed re-creations all the time. For de la Peña, careful attention will need to be paid to good research and transparency in how
scenes are depicted, but the benefit of VR is too important to dismiss due to worries about potential abuse.

VR's emotional power is apparent in *Kiya*, a five-minute-long
VR experience that depicts the moments leading up to a fatal
domestic murder/suicide. Viewers stand inside the home with the
woman's gun-wielding ex-boyfriend, who holds her hostage while
her two sisters plead with him to let her go. Though the avatars look
a bit cartoony due to the commercial-grade computer graphics,
the audio—captured from two calls to emergency services—leaves
no doubt about the underlying reality of what we're observing;
it is chilling when the scene reaches its terrible climax. At that
moment, de la Peña makes the decision to take viewers outside of
the home, where they witness police officers approach the suspect
in an attempt to defuse the situation. Gunshots are heard, and
the piece comes to an end. The decision not to depict the deaths—
a decision Morris also made in *The Thin Blue Line*—shows a
welcome restraint and respect for the reality of the events being
depicted. But in the increasingly hypercompetitive profession that
gave us the phrase "If it bleeds it leads," I am not optimistic that
restraint and respect will always win the day. How to deal with
this kind of real-world violence in VR journalism, and what exposure to violence and death in VR will mean for viewers, is a concern
that future standards and practices in VR will have to address.

In 2016 the Journalism Department at Stanford University
sought to study these types of issues in a class dedicated to immersive journalism—one of the first of its kind at any university, to
my knowledge. During the 10-week course, 12 undergraduate and
graduate students evaluated a range of virtual reality experiences
published by the *New York Times*, the *Wall Street Journal*, ABC

News, and others, focusing on one overriding question: Why use virtual reality in your storytelling? In what situations does it add to a journalistic narrative? The answer, they determined, was not as often as you might think. The conclusion of the class was that at the moment VR added value to the story in only specific cases, and then only as a supplement to a traditional reporting style. Most VR news stories are done through immersive video, and capturing this video can be complicated in a journalistic setting. For one, because the camera is recording images in 360 degrees, if the camera operator doesn't want to be confused with a character in the scene she is filming, she must run out of the spherical shot after setting it up. This means that which events are captured by the camera will be happening outside of the photographer's control, and perhaps without her knowledge. This does not lead to the most riveting footage. It can be difficult to tell a story—an act of direction and selection—when the camera eye sits in one place passively.

In a way, our early VR journalists are in the same situation the nineteenth-century photojournalists were in when they took their cameras into the field. Hauling around heavy, delicate equipment that takes a long time to set up and to execute a shot, the best way to ensure a good product is to set the scene. This is why some of the earliest experiments in VR journalism are documentaries with staged shots, dramatizing the lives of the people depicted, or static scenes of organized activities, like demonstrations, vigils, and political rallies. It may be some time before VR photojournalists are able to venture out into a breaking news event and capture effective footage in the middle of conflict.

Situations in which the surroundings or environment are a crucial part of the story make for the best VR, as the best, early practitioners are starting to realize. When I spoke with her, de la Peña described an editorial piece she was planning about the increasing prevalence of air rage incidents. Her VR experience would put the viewer in an airline seat from decades ago, and then watch as the comfortable dimensions of that environment around the user slowly begin to shrink, as more and more seats and people are added to airplanes in the ensuing decades. Why is there an increase in air rage? Because we're being crammed into planes like sardines in a can—a concept that is brilliantly illustrated around your virtually embodied self. Another good example of this type of storytelling is *6x9*, a VR experience produced by the *Guardian*, which takes users inside a solitary confinement jail cell. Claustrophobia is one of the tried and true parlor tricks in VR.

What makes each of these examples a good use of VR is that the entire environment surrounding the viewer is important to the story, and requires the viewer to explore it by turning his head or body. If all the action is front and center—say at a political debate—you don't need spherical video. There is little point in filming in VR if the engagement with the story only requires you to look in one direction.

All these early experiments in virtual journalism raise the urgent question: If media organizations build VR, will the audience come? Will VR stick or will it go the way of 3D TV? They share some of the same issues—expensive technology and goofy, uncomfortable, wrap-around glasses. But there is too much investment and development going on in the space—it is already likely too big to fail. How is VR different from 3D TV? There was never a clear content reason to invest in 3D TV. No "killer" experience emerged, so content producers never reached the critical mass of the market. With the exciting content that has been produced so far, the trajectory for quality content in the VR space already has a foundation. Once you experience a VR "aha" moment, you can't wait to find the next one.

Ethical and practical considerations will be at the center of journalistic VR's development in the coming years, but aesthetic matters about *how* to tell stories in VR, and what techniques bring about the strongest emotional effect on the viewer, are unlikely to develop much in the news business. As with documentary films, which by and large adopt a straight narrative style, nonfiction narratives are generally focused on conveying information. While affecting viewers emotionally can be an aim for journalists, it is not usually the point of their work—in fact, journalism's preference for objectivity argues against overly emotional reporting.

But what about narratives that are designed to create emotional reactions, that are unbound by the restrictions that guide news stories? An industry is already growing in Hollywood and Silicon Valley to explore VR as a space for fictional narratives, and within it storytellers from Hollywood and the gaming world, with technical help from technology companies, are beginning to take the tentative early steps in defining the grammar of virtual storytelling.

Brett Leonard was a young filmmaker fresh from his hometown of Toledo, Ohio when he landed in Santa Cruz just before the beginning of the first VR boom in 1979. There he fell in with future Silicon Valley icons like Steve Wozniak, Steve Jobs, and Jaron Lanier. Jaron is also one of our most incisive and visionary thinkers about technology and its effects on human commerce and culture; at the time, as well as now, he was the very public face of virtual reality, a term he coined and popularized. Leonard was a big fan of technology and science fiction when his trip to Silicon Valley dropped him into what must have seemed the most interesting place in the world, amidst a group of people who were already playing a significant role in shaping the future. It was in this environment of optimism and excitement that he began sampling some of the early VR prototypes at Lanier's company, VPL, and engaging in long chats with Jaron about the creative possibilities of VR as a medium for art and expression.

There is a direct line from Leonard's introduction to VR and the first feature film he directed, *The Lawnmower Man,* a successful 1992 independent movie that's a cult classic for many VR fans. Cowritten by Leonard and only nominally based on the Stephen King short story of the same name, the film told the story of a scientist who uses VR to enhance the mind of a mentally disabled man, unwittingly turning him into a diabolical, violent genius. As Leonard describes it, "It was Mary Shelly's *Frankenstein* at its core, with a little bit of Daniel Keyes's *Flowers for Algernon* thrown in there, and *The Sixth Finger* episode of *Outer Limits.*" Many of the big VR themes are in the movie, including fears about behavioral modification, addiction, and the consequences to real-life relationships of absorption in digital worlds. But the movie also puts forth a positive vision of VR technology as a powerful force for training cognitive skills and encouraging creativity . . . if used wisely. VR may be "the most transformative medium in the history of humanity," the movie warns, or "the greatest form of mind control ever devised." This idea, articulated a quarter-century ago, still defines Leonard's thinking about VR today.

After decades of directing films and TV, Leonard has returned to VR, forming a content studio with a multidisciplinary group of creators called Virtuosity. In late 2016 he visited my lab to learn about our research, and I had a chance to talk with him about the future of storytelling in VR, a subject to which he's given a lot of thought in recent years. Like many of the storytellers who are trying to figure out how VR might work for fictional narratives, he draws on the development of film to guide his thinking about VR's future. One lesson that clearly emerges from that history is that it takes a long time—generations—for artists to exploit all of the possibilities of a new multidisciplinary art form, and that these artistic possibilities can be influenced as much by business and technological developments as they are by conceptual breakthroughs.

To paraphrase one of Marshall McLuhan's most significant insights: people using a new medium have a difficult time breaking out of the thinking involved with the previous ones. We see this in the history of Hollywood filmmaking. Many of the early storytellers in Hollywood came from the world of the stage. Consequently, early directors essentially filmed stage shows—one camera angle in front of a proscenium arch, with few to no cuts. It wasn't just the directors: early performers, not used to the unique intimacy of film, made broad motions when they acted, as if they were trying to communicate with a person in the back row, and not the camera five feet from them. Over those early years, directors and performers quickly innovated their approach to filmed stories, the former introducing editing and camera effects that began to reveal how strange and surreal this new form of storytelling could be. By the '20s, cameras were much more mobile, and cuts were more frequent between shots that were now coming from a variety of different angles and levels of focus. You can see these techniques on display in the often frenetic slapstick comedies of Buster Keaton and Charlie Chaplin. But they could be seen in dramatic pictures, too. The early German expressionist horror film, *Nosferatu*, uses lighting and editing techniques that emphasize the visual nature of film. Of course, that was the silent era, but the introduction of sound in the late '20s shifted film to a more narrative, story-based medium, and the content shifted with it.[7] As this brief description of the changes in early films shows, there isn't necessarily a linear progression, or a logical "end" to what a given art form becomes.

Instead there is a constant mutation influenced by what technological possibilities allow it to be, what creators want to do with it, and what audiences are willing to pay for.

The influence of market on what gets made can't be overemphasized, and it is an important element to what will happen to VR in these fledgling years. We can expect to see a lot of experiments and a lot of significant changes as content creators work out what people want. Leonard points out that the earliest commercial motion pictures were nickelodeons, brief silent films that could be viewed in machines in popular public places like boardwalks. For some pocket change, viewers could get a few minutes of this exciting new technology in action, and see performances on demand that previously would require a theatre ticket. After nickelodeons came one-reelers, which lasted about 10 to 12 minutes. Longer-running feature films didn't become popular until D. W. Griffith's *Birth of a Nation* in 1915. "People laughed at him and called him nuts," Leonard told me. "People thought 'no one's going to sit still for a movie for over 20 minutes.' Everything they're saying about VR right now was being said at that point when the feature film was becoming an actual product." I laughed when he told me this, given the 20-minute rule for wearing the HMD in my lab today.

That's an important point to keep in mind as we speculate about what VR storytelling might look like in the next 50 to 100 years. What we call "the movies" today has been a constantly changing medium since its beginnings, the product of market forces, individual artistic genius, and constant studio investment in technologies to create new, more thrilling experiences for movie-lovers. Sound, color, steady-cams, surround sound, 3D, IMAX, Cinerama, digital recording—these are just a small sample of the changes that have altered the palette of options filmmakers can explore in creating stories.

Where are we today with VR? Even the most optimistic content creators working in the medium today admit we are closer to the nineteenth-century experiments being performed by film pioneers like the Lumiere brothers than we are to Orson Welles's *Citizen Kane*. Leonard points out that the early attention paid to basic VR experiences like Google cardboard, while it has helped reach potential users, has undermined the possibilities of VR by demonstrating an inferior version of VR that is unlikely to excite people. "This is the beginning of a new medium," Leonard pointed out. "We've got to have a process that creates R&D and a product at the same time. That prototyping R&D/product-making is the moment we're in. I would say the biggest part of this moment is actually making the market."

"Making the market," to Leonard's mind, is going to require a substantial investment in talent and technology. He described how MGM took a huge risk in creating *The Wizard of Oz*, the movie that is arguably the precursor to the modern special-effects-laden spectacle film that dominates our domestic and global markets today. "MGM said we're going to make these big cranes, and these dollies, and all of these incredible tools. We're going to create a journeyman's process for actually training craftsmen. They created the idea of the spectacle film. That is still at the core of what the business is today. That was a very important thing that was not easy to do. They had to embrace all these different disciplines, and bring them together under one roof." This raises the question, who will be the MGM for VR entertainment?

Perhaps the most recent example of the big-risk, paradigm-shifting Hollywood movie is *Avatar*, James Cameron's sci-fi epic, which pioneered new advances in digital effects and 3D filmmaking that have changed the way movies get made. The head of the visual effects team on the film was Robert Stromberg, and it was in 2006, during the making of the film, that he got his first taste of VR—even though he hadn't put on a headset. "We were doing something unique," Stromberg told me, "in that we were creating 360-degree worlds in a CG world and essentially virtually photographing those worlds with a virtual camera. It's the first time that a film had ever been attempted or been made in that manner, and I was on that film for four-and-a-half years. What it did was it opened my eyes to something new, and that was the ability to transport yourself into a virtual space."

Stromberg went on to work on the production design of other films, including *Alice in Wonderland* and *Oz The Great and Powerful* and *BFG*. He then made his directorial debut with the movie *Maleficent*. But even as his career in Hollywood was thriving, he kept his eye on developments in VR technology.

Then he read an article about the purchase of Oculus by Facebook in 2014. "I literally cold-called Oculus that very day the article came out and asked if I could come to see what they were developing," he told me. "To my astonishment, they said yes. They were a small company in Irvine, California at the time. It was really a room full of techs and relatively unimpressive." He ran through two demos that day. One showed him in a room with robots of different sizes that illustrated how presence could be created by being able to move your head and look around objects.

The second experience began in a small room that gradually grew bigger. As the walls receded he was left feeling as if he were floating through this changing, metallic world. "That's when it really hit me," Stromberg said. "I'd spent my entire life creating these worlds, and for the first time, I felt like I was able to actually step into it." That day, Stromberg called some friends and started the Virtual Reality Company. He knew that the tools were finally there to make virtual reality experiences available to a mass market. But what no one was really doing yet was making content.

Stromberg's first experience was called "There," a four-minute test that showed a dream landscape containing a series of islands floating in a void, unbound by the laws of the physical world. A young girl leads the user through this surreal terrain as a cinematic score plays in the background. Stromberg wanted some feedback, so he took this brief experience to Stephen Spielberg at his house on Long Island and showed it to him. Spielberg was blown away. Spielberg shared it with his grandkids and his wife, the actress Kate Capshaw. "She literally had tears after she watched it." Validated by the enthusiasm of Spielberg, who came onboard as an advisor to VRC, Stromberg was even more confident that he was dealing with a medium that could convey the magic and emotion that storytellers are looking for. "I could see the path forward."

The intervening years have been a process of continued experimentation for Stromberg and his colleagues at VRC as they attempt to address the many hurdles and take advantage of the special opportunities that VR presents the storyteller. Perhaps the biggest difficulty for a traditional filmmaker is to negotiate the dictatorial demands of storytelling—the imposition of an artistic vision—with the interactive aspects of a medium that gives you the freedom to look anywhere you want, and potentially, to move around a scene. In VR, there are no constraints on attention; it's utter anarchy for the senses. Movies are the opposite—at any given point, you only see a tiny fraction of the world—the small window inside the camera's field of view. It's impossible to turn around and look behind the camera. Or up at the sky. Imagine any classic movie scene that focuses on a character's face. In that particular point on the timeline, if you wanted to rotate around and look to see just what the character is looking at, you couldn't. And for good reason, the director chose that moment to show you the hero's face; that framing is an essential point of the narrative. If you were turned around you would miss the subtle glint in our hero's eye that makes the scene so special.

VR is a democracy. You can look wherever you want, whenever you want. In film, the director is a dictator, one who controls your senses, and forces you to look where she wants, when she wants you to. Movies succeed as a medium because directors are quite good at this.

One of the most basic concepts in storytelling is something called "plant and payoff." In one scene, the author or director plants the seed of a twist that occurs later on. Maybe it's a side-long glance, or a wallet sitting on a table. Sometime it's done with dialogue, other times with just imagery. But it's typically subtle. To take an example from the classic film *The Shawshank Redemption,* it's the seemingly incidental images of a rock pick and a Rita Hayworth poster. We see them every once in a while in the film, but it's a very small portion of our field of view. The payoff—that the hero uses the rock pick to dig a tunnel, and then cover that tunnel with the poster—is the high point of the film. It is essential for directors to make the plant memorable enough to be noticed, but not so obvious that the audience can predict the twist later on. This general theme doesn't work well in VR—if it's subtle, it is easy to miss. Instead of seeing the subtle rock pick, the viewer might be looking off into the distance at the other jail cells. Or at the ceiling, marveling at how real the lighting effects are. VR is typically pretty interesting, and attention tends to wander.

VR is about exploration, and storytelling is about control. This conflict has led to two major strategies in VR. The first, one that you see most often in 360 video such as the *New York Times* VR application and other journalism VR stories, is to basically put the action all in one spot. There is a person talking whom you need to look at, or a main view of a scene, and while sometimes people do look around, for the most part people look straight ahead of VR. This of course raises the question, "Why would you do this in VR?" If you are looking in one direction anyhow, you might as well just look at a TV screen. Indeed, a sure way to detect VR experiences that take this strategy is by watching the people who are wearing the goggles. If they are just looking straight ahead the whole time, then likely the content resembles that of a traditional film that happens to be centered in a sphere.

The second uses all of the space in the 360 sphere, and there is critical action everywhere. If it's "room scale" VR, then the people are walking around, looking around, and they are completely leveraging what makes VR special. But more often than not, they are highly engaged by the experience, yet they aren't hearing the story the director wants to tell.

VR filmmakers are trying a variety of techniques to balance exploration with storytelling, and it's telling that the creative minds working on a lot of early VR stories come from both the world of videogames, which feature interactivity and exploration, and the world of cinema. One strategy is to use sound, movement, and lighting cues to bring the gaze of the viewer toward the action the director wants them to see. (A nonentertainment application, like a training video, might use arrows to do the same thing, but this solution obviously won't work in a narrative that wants us to suspend our disbelief.)

The most common solution is to use "spatialized sound." Basically, you can beam sound at different volumes to each ear in order to make sound seem as if it is coming from various places around your head. Most of us have experienced something similar at a movie theatre that is designed to have great surround sound. This strategy is less obvious and therefore less effective at turning heads than the arrows, but assuming the sound is integrated into the theme of the narrative, this trick at least doesn't break presence as much as a dancing arrow.

Another possible solution is to have the action in a scene pause, or not progress, until the viewer turns her head to exactly the right spot, at which point the critical action can occur. The "plant" waits until the viewer is looking, then proceeds to play itself in order to ensure the payoff makes sense later on. So with the *Shawshank Redemption* example, the movie holds on pause until you look directly at the rock hammer. After a few seconds to ensure you have noticed it, the movie proceeds. As you think through that one, you'll realize that these movies would have no set length—90 minutes for some, 180 for others who take longer finding the hotspots.

Another problem is narration. VR storytelling tends to be heavy on it. The problem is people don't want to listen to monologues— they are in virtual reality, which is a perceptually rich and fascinating experience. But storytelling, as it exists in most media, involves talking, and even worse for VR, long monologues. So the viewer is typically given a choice—dedicate the lion's share of her attention to listening to the words (and miss a lot of the action going on around her), or turn around and gaze at all the amazing visuals, but don't hear the narration with the level of attention a director would hope for. My lab at Stanford has personal experience with this conundrum. In one of our early versions of "The Stanford Ocean Acidification Experience," our field trip to the bottom of the ocean to teach marine science, we had a constant stream of narration and a constant stream of visual action.

People chose to look—they reached out at colorful fish, marveled at the coral around them, and were very high-presence in the underwater scene. Unfortunately, in that first study it meant the science—which was delivered via narration—fell on deaf ears. The experience was a hit, the narrative an afterthought. In later versions we have been careful to alternate great narration and great visual action so that they don't compete with one another.

By now a picture should be emerging. VR is great for experience— it is organic, user-driven, and different for everyone. Film and prose are great for telling stories—where your attention is guided nonstop by a director or writer. How these different domains will be integrated, and what that will mean for our traditional idea of storytelling, remains to be seen.

"The rule book is still being written," Stromberg told me. "By no means am I comparing VR to film, and by no means am I comparing it to a stage play. I think it's a hybrid of a lot of things. The traditional sense of editing and framing are out the window." Stromberg returned to the idea of scale that so impressed him on his visit to the Oculus offices. Watching a scene in VR, he pointed out, "You feel like you're standing there." Shifts in point of view in VR don't feel like interesting disembodied new perspectives. Instead, they feel like your body has changed. When a filmmaker does a close-up in a traditional movie, for instance, it doesn't feel as if you've invaded the actor's personal space. But in VR it does. "As a director you must be cognizant of space and scale more than you would in film. It's just a different way to approach a story."

Perhaps a more fundamental question about VR storytelling is how people will prefer to view these stories. Will they want to observe narratives happening as a kind of ghost, moving around the scene but not actually involved? Or will they want to be embodied actors participating in a plot that reacts to their actions? Right now most of the creative talent in VR film is coming from Hollywood and the videogame industry. The expertise of the folks from Hollywood is with the former version of narrative and those from videogames with the latter. The engagement that comes from interactive narrative, and the promise of exploration, deeper engagement, and repeat use are all strengths of emergent storytelling. But what is lacking in almost all but the most exemplary game narratives is true emotion and efficient narrative structure. For instance, well-executed examples of traditional narrative forms, in which the experience is guided by a writer or director, are carefully paced. Game narratives, which allow the player to wander about and explore environments in between narrative beats, do not generally result in economical storytelling. In the near future, there will most likely be room for both, but it is the possibility of large social shared narratives that excites Leonard as he thinks about the future of immersive entertainment.

What does Leonard imagine VR storytelling to be? Like everyone working in this space, he'll admit he has no idea, but he's sure it's going to be profoundly different from film, which has seemed to run out of new ideas in recent years, if the big-budget offerings of Hollywood are any indication. "That's one of the main reasons I'm in VR—it has a profound opportunity to really explore imagination again. You don't have to go in and basically get something with a guy in a spandex suit and a cape, because that's the only thing you're going to get traction with."

VR, Leonard believes, will explore an undiscovered country between gameplay and linear narrative. "In cinematic storytelling, you've got story, character, and emotion. There's the linear journey there. Plot and story is very structural. It brings you into it. The screenplays are very focused on that." But VR, he believes, will be driven by emotions and character, and then discovery of story. It won't be storytelling, but a process he calls "storyworlding."

"Once you start thinking about that, and start building a world from its base elements up, you start realizing that you discover narrative. The discovery narrative has to be embedded in the process of the group creating it so that it can be endemic and organic to the process of participants experiencing it." Leonard sees future, narrative-rich entertainment spaces in which stories can be launched through the use of what he calls "narrative magnets," the way a mission starts in a videogame when you walk to a certain point on the map. With good AI, the narratives and stories would emerge from the interactions of individuals.

Leonard imagines a group of friends in a story world, each playing a different role in the drama. Perhaps it's a spy narrative set around a baccarat table in a luxury casino. Maybe someone is playing a waiter, someone is the spy, and someone else is the croupier. The action starts when everyone is in place. Perhaps a James Bond figure's nemesis, portrayed by a virtual human with artificial intelligence, steps to the table and lays down a big bet. The action would then unfold, with each character witnessing events from a different vantage point and acting as he sees fit. Maybe your character, a tourist at the hotel, finds herself roped into an escape plan for the spy—his love interest grabs you by the hand and leads you into a back room. Now you're engaged in a chase scene with bad guys. Such a scene could be a side mission on a longer story line for your character.

"I think linear narrative is a clothesline, and there are all these clothes hanging on the clothesline. You can take them off. You can put those clothes on." After deviating from the main story line you snap back into it after this escapade. Leonard believes the need for resolution is an essential part of storytelling and that these longer story arcs will be character-specific, but all sorts of entertainment possibilities can be offered in a world that has been designed by storytellers.

Of course, VR entertainment won't be one thing, just as filmed entertainment is not one thing. It is music videos, documentaries, feature films, animated shorts, 3D IMAX experiences, and everything in between. There is every reason to believe that VR will offer a similarly expansive array of entertainments, accessible not only from home but in arcade settings. Already there is The Void, a VR entertainment space housed in Utah that allows groups to experience social VR scenarios as they walk through real-world spaces outfitted with a variety of haptic feedbacks that correspond to the VR game world. Planned scenarios involve exploring an ancient tomb, battling science-fiction aliens, or even fighting paranormal enemies as a Ghostbuster.

The funny thing is, looking back to the beginning of VR, I don't think many people thought it would be a way to hear stories. In my last book, *Infinite Reality*, we had a chapter on VR Applications, and we didn't even touch on the subject. I can't speak for my esteemed coauthor Jim Blascovich, but it never would have occurred to me to have a chapter about movies or news pieces. When we think of the pioneers of VR from decades ago—Jaron Lanier, Ivan Sutherland, Tom Furness—nobody was talking about storytelling, or at least not as a centerpiece of their vision. It just didn't seem like an obvious use case, likely because of the constraints we mention above. But the film and news industry are betting, heavily, that this medium is their future. I am skeptical about a lot of it.

Even if the narrative issues get resolved, there is a disconnect between seeing and doing. I love to watch zombies on TV—it's a guilty pleasure of mine. Any George Romero movie, *The Walking*

Dead, you name it. But I would not want to kill zombies in VR, or have my flesh torn off my arm by rotted teeth. Imagine the first-person point of view, the haptic effects, even virtual scent. I've seen some preliminary zombie VR demonstrations secondhand. I have yet to put on the goggles and go in myself. Movies seem to have just the right amount of engagement—viewers are captivated by them, and almost by definition a good movie causes the viewer to connect with and take the perspective of the protagonist. But psychologically, not perceptually.

Even so, sometimes a movie causes nightmares. Certainly, the movie *Psycho* caused people to look twice in the shower, and *Jaws* might have created some phobias of sharks and oceans. But the difference between seeing and doing is massive. Every chapter in this book is written to support the argument that the brain treats a VR experience in ways similar to a real one. Think about some of your favorite movies, and imagine the scenes actually happening to you. Quentin Tarantino would be out of business in a heartbeat.

CHAPTER 9

REVERSE FIELD TRIPS

When I was a kid in the 1970s, there was not much pro-gram-
ming on television aimed at kids, and most of what was available wasn't very enriching. In fact, in my house, "screen"
media pretty much consisted of football games, *All in the Family*, *Barney Miller*, and the occasional trip to the movies. The one
exception to this was *Sesame Street*, which got its hooks into me
when I was three years old. I loved the characters and I loved the
urban setting. Though I didn't live far away from New York City,
the city streets depicted on the show seemed as exotic to me as
another planet. Amidst all of the excitement, I barely even realized
that the show was teaching me anything.

I also had no idea that I was participating in a great social experiment. At the time I was watching the show, the idea of using television as an entertainment and educational tool was pretty new, and
controversial. It had its roots in mid-twentieth-century advances
in child psychology and development, but those ideas didn't carry
over into television until the late '60s, when *Sesame Street* first
aired. As I said, there wasn't actually much on television for kids
at all—we weren't much of a consumer market, and parents of that

generation were as a whole, let's say, less persuadable by entreat-
ies for new toys. (Which is understandable because *their* parents
grew up during the Great Depression.) In any case, at that time,
television programming was dominated by still-familiar genres
like soap operas, sit-coms, Westerns, and live sports. As an
advertising-driven medium geared for mass-market consumption, television was not considered particularly enriching, and
was maligned by sober-minded critics who derisively called it the
"boob tube" or "idiot box." But in the '60s, the pioneering television producer Joan Cooney saw in the medium's absorbing nature
an opportunity to teach young children the skills they would need
to succeed in school. In 1968, Cooney founded the Children's Television Workshop (CTW, known since 2000 as Sesame Workshop)
and staffed it with a host of child-development psychologists who
rigorously studied its effectiveness in capturing the attention of
children and imparting knowledge.[1]

One of those early researchers was Lewis Bernstein, who first
saw *Sesame Street* in 1970 while getting his Master's in psychology
at Hebrew University in Israel. At the time, he was becoming disen-
chanted with the theoretical focus of the curriculum in Israel, and
wanted to put his research work into action to help disadvantaged
children reach their full potential. When he saw *Sesame Street*, he
was struck not only by its ambitious efforts to teach cognitive skills
to children, but also its dedication to imparting social, moral, and
affective lessons.[2] Monsters of different colors lived (mostly) in
harmony in the show, and the humans that interacted with them
represented individuals on the entire spectrum of American
life—adults and children of all races, be they rich or poor, urban
or rural. "They had," Bernstein told me, "created a community
in which people were supported and nurtured, and there was joy
in learning." After completing his studies, Bernstein returned to

his home in New York City and approached the CTW to see if he could work as an intern. Nothing was available, but after an interview with CTW's head researcher, Edward Palmer, he was quickly offered a full-time job once they realized his rigorous education in Israel meant he'd already read everything published on the subject of childhood development. Bernstein would stay with CTW for over four decades, working in a variety of roles from research to executive producer.

While the show was designed to appeal to all children, it was a particular goal of the show's creators to reach inner-city children and low-income households, and to use any new medium that was appropriate for learning to enrich the educational opportunities of the disadvantaged. That's how in 2013 I had the privilege to begin working with Bernstein, who was looking to explore VR as a platform for *Sesame Street*. At the time, he was vice president of education, research, and outreach. We began collaborating on a series of experiments examining children's use of VR, and during that time I got to hear many wonderful stories about the history of the show. Over lunch in 2014, while we were planning a research study together, I began enthusiastically telling him about my favorite memories from watching *Sesame Street*, particularly my favorite segments in the show.

Most kids are obsessed with *Sesame Street*'s charismatic monsters like Big Bird, Grover, and Oscar the Grouch. (We didn't have Elmo when I was a kid.) But I really loved the field trips, the moments in the show when I got to tag along with other kids to places I couldn't see in my hometown: museums, science labs, the insides of farms, factories, and dance studios. Bernstein was delighted to hear it. While the skits are loved by children, he felt that the field trips—which were a very important part of *Sesame Street*'s educational vision—are often overlooked because they are not fantastical, but real. But the CTW wanted not only to teach kids their 1, 2, 3s and X, Y, Zs—they wanted to give them a sense of the world's size and diversity. For a kid in the country, that might mean seeing a skyscraper in New York City or watching city kids play stickball on a street in Brooklyn. For the urban kids, it might mean seeing a dairy farm or visiting a state fair. There wasn't an urgency in these segments—events would unfold slowly and details would be explored by the camera. Putting the viewer in these exotic settings opened up new imaginative possibilities—especially for children from low-income households who do not get to travel often. It also enlarged the empathic imaginations of everyone who was watching, widening their understanding of the varied lives of people in different communities.

Today, there are multiple channels on television that show cartoons 24 hours a day, seven days a week, not to mention on-demand streaming services and a seemingly unlimited supply of children's movies available at the press of a button on screens of all sizes. Many of these shows are good, and almost all of them are slick and polished compared to children's programming in my time. But there is something worryingly frenetic about the pace of these shows. In today's media marketplace, a different kind of content and tone is prevailing; and like many older media institutions, *Sesame Street* is struggling to compete. One of the main conversations at *Sesame* is how to make the show more competitive. Sadly, it turns out, one answer to that is to reduce the number of field trips. Kids *love* cartoons. But trips to the science labs and factories, not as much, even if you put a loveable character like Elmo on the scene. Nothing beats cartoons. This was a trend Bernstein was ruefully observing as his four-decade-plus career at *Sesame Street* was winding down. He left with a strong plea to keep the field trips in. For kids, especially those who don't get to travel often, real video of other kids going to special places was essential to learning and preparing them for school and beyond.

The field trip is the perfect metaphor for VR learning. On a field trip, you get to go somewhere special, a place where physically being there matters. For example, growing up in upstate New York, we used to take nature walks to Piano Mountain. They were led by a naturalist who would point out trees, birds, and salamanders. But there was something special about being there—we were out of the classroom, and it was truly a teachable moment.

You don't go on field trips every day, of course—they are designed to augment the classroom, not replace it. VR should be the same. As my colleague Dan Schwartz, a cognitive psychologist turned dean of the School of Education here at Stanford, likes to say, even though there is a lot of emphasis on learning by doing—so-called experiential learning—the "telling" still works pretty well.[3] For most students, most of their college educational experience still involves listening to a teacher give a lecture. It's a default routine for educational scholars in promoting experiential learning, but the truth is, we learn a lot from simply listening.

As we learned in the chapter on social VR, there are still some hurdles to overcome before VR can effectively mimic the advantages of learning in a classroom. I think we are still some years away from that. But the ability to immerse a person in another environment—one that will impart knowledge and open up a person's mind to the richness and variety of communities and natural phenomena beyond his or her reach—is a powerful tool. And it is already possible. VR field trips are ready now.

Harvard professor Chris Dede, who has been creating virtual-reality learning scenarios for more than 15 years, is a trailblazing researcher in this area. In his 2009 landmark article published in *Science*, Dede outlined all of the pedagogical advantages of immersing a student in a VR field trip.[4] Dede's VR field trips have all of the assets of an actual field trip—going to special places specifically chosen for teachable moments—but they also allow for the impossible. Imagine visiting an ancient ruin and then traveling back in time to see it restored to its former glory. Or reliving teachable moments with different choices, turning physics on and off, or experiencing a pivotal moment in history from different points of view.

Across years of research, scholars on Dede's team have shown VR helps learning. His *River City Project*, a multi-user virtual environment (MUVE), created an interactive simulation of a nineteenth-century town that allowed middle-grade students to apply modern knowledge and skills toward a host of medical problems facing the virtual inhabitants. Dede was able to demonstrate that students who were asked to confront a medical science dilemma, such as an outbreak of illness, had learned more about epidemiology and prevention than they had in traditional classroom settings. In particular, the immersion experience inspires learners to spend more time learning. They are simply more motivated to attend to the material. This is especially the case for students who typically have difficulty in the classroom. Dede's field experiments from schools show that VR is a particularly effective tool for students at risk, especially those who could never imagine themselves doing science. By taking the role of a scientist in the virtual world, they gain self-efficacy—the belief that they actually *can* do science. In the words of Dede and colleagues:

"Thousands of students and hundreds of teachers around the US and in Canada have used *River City*. Our research with River

City has found that MUVEs increase engagement in learning by allowing students to immerse themselves in a virtual world. River City has been particularly effective in motivating students who are usually unengaged and low-performing academically. Controlled studies show educational gains in science content, sophisticated inquiry skills, motivation to learn science, and self-efficacy."[5]

I have worked with Dede on a few projects. One thing that became clear through our collaborations is how much time, effort, and money went into building his VR Field Trips. Years of blood, sweat, and tears from teams of engineers and programmers, 3D artists, education specialists, story-boarders, and actors. It is no small effort to create an engaging, interactive, and most importantly, scientifically authentic field trip—much less one that is teacher-approved. Most VR designers working today are building content that captivates a user for a few minutes at best, and even those don't come cheap. For example, the production costs for the 360 videos that are being produced by news organizations like the *New York Times* can exceed hundreds of thousands of dollars for a small set of videos—and they aren't even interactive. Dede's *River City* and *Eco-Muve*—his two most ambitious experiences—can engage learners for hours, even days. I have been in the field of VR for two decades, and the educational field trip is the elusive unicorn. Everyone talks about how it's the greatest use of VR—but almost nobody has a good example to show you. To quote a colleague who was referencing VR in the corporate world, it's like having sex in high school—everyone is talking about it but nobody is actually doing it.

So the bad news is, VR field trips are a massive undertaking to build right.

But the good news is that once they are built, they can be distributed on a massive scale, enabling us to share educational opportunities with anyone who has an Internet connection and an HMD.

Just as YouTube instructional videos have provided free lessons to people all over the planet, in the near future VR users will be able to access educational immersive environments from anywhere, at any time. Massive Open Online Courses (MOOCs) showed a similar promise of disruption—once a lecture is stored it is viewed by people all over the planet. For example, Andrew Ng, Stanford professor and chief science officer of the massive Chinese technology company Baidu, has produced the canonical MOOC, "Machine Learning." In some ways, it is the MOOC that started the online education revolution. That class has a total enrollment of well over a million students from all over the world.

A graduate student who did his dissertation work in my lab, Brian Perone, ran what is likely the first immersive virtual reality study in high school classrooms. He chose a local high school in the Palo Alto area that allowed him to set up a full VR system in the room and use it to share the VR field trip about ocean acidification described in Chapter 4. We created an active lesson on how oceans suffer due to CO_2 absorption, where students became scuba divers and went to the bottom of the ocean. Here is the amazing thing about his study— he had a perfect control condition: the students from his class also went scuba diving for real, on a real field trip to Monterey Bay.

We'll discuss the results of his study later, but for now I want to focus on what an utterly amazing opportunity these students had. Most readers cannot remotely fathom getting taught how to scuba dive in high school. Diving is ridiculously expensive, and many high schools can hardly afford textbooks and Scotch tape. These students were born into an economic situation that allowed them to have a learning experience that is available to far less than 1% of students in the world.

But copying and pasting digital field trips is free. Once we build one, we can build a billion. So just as Andrew Ng has taught over a million people how to build neural networks and support vector machines, future students will be able not just to hear any lecture, but to go on the most expensive, rare, dangerous, or even impossible journeys. Virtual scuba diving doesn't require certification, insurance, fancy equipment, or gas money.

But here is the million-dollar question—how educationally effective are field trips in VR? What are the design principles that should guide these types of experiences?

As we have seen, VR works quite well for training very specific skills, whether it is athletes using STRIVR, surgeons practicing laparoscopic surgery, or soldiers training in a flight simulator. Some of the most famous training work, done by USC engineer Jeff Rickel in the '90s, has demonstrated that VR training works for examining engines on large ships. But that is a very different use case than teaching science and math—so called STEM learning— which is a less procedural and more cognitive endeavor. For science learning, as we discussed earlier, Dede has for years been showing improvements in science test scores with his *River City* and *Eco-Muve*. But Dede's system is "desktop" VR, not immersive VR using a headset. It is more of an interactive videogame than the true VR we describe in this book.

In immersive VR, where people are fully present inside pedagogical scenes, we've seen a number of studies that show that VR field trips facilitate learning. For example, when Brian Perone looked at the effectiveness of VR ocean field trips in high school and college classrooms, he found a large knowledge gain through the virtual field trip when he examined test scores from before and after the lesson.

But the million-dollar question remains: Given the cost of building great VR simulations—not to mention, as Perone found in his high school setting, a tendency for kids to tease one another as they grasp at fish that aren't there inside the headset—is it worth teaching lessons in VR? Wouldn't watching a video suffice? Or just reading a textbook the old-fashioned way? There is still a lack of data on the topic.

One of the first rigorous studies to compare immersive VR learning to learning from a traditional computer screen was conducted in 2001 by my colleagues at UCSB during my postdoc there. Richard Mayer and Roxanna Moreno examined the effect of immersion on learning. They created a virtual world to teach participants about botany, and had some participants experience the world via headsets while others viewed a computer screen. In particular, they looked at two classes of learning. The first was retention, a simple memory measure for the facts presented about how plants live and grow. In general, education scholars are less concerned about retention and more concerned about transfer—the ability to take the lessons learned and to apply them to novel situations. For example, in the transfer test questions, students might be asked to "Design a plant to live in an environment that has low temperature and high water table." While the facts they learned in the virtual world allowed them to infer a proper answer to the new question, the exact details were not provided directly in the learning materials. Mayer and Moreno found that students were more engaged in the more immersive headset compared to the desktop computer. However, their research showed that engagement did not increase learning. VR was more immersive, but didn't increase test scores.[6]

In my lab, we have run a handful of similar studies in which immersive VR is compared to desktop and video in teaching a science-based lesson. In just about every study we run, we demonstrate an increase in knowledge acquired via the immersive VR lesson. Students clearly learn about science in VR when you test them both before and afterward. But the story becomes more complicated when you compare learning in immersive VR to less immersive systems. Typically, we find that VR produces attitude change—people will care more about the lesson topics and will more likely agree with the point of view of the lesson. But we've found small if any changes in the retention of facts. This has always been puzzling.

Dede argues persuasively that VR should cause more learning transfer than other media, due to the ability of students using VR to take multiple perspectives in a scene and because students can learn in a context that feels like the real world. Given that the instruction is delivered in a complex and interactive situation—for example, a city trying to understand an outbreak—the ability to apply the information to new contexts should be enhanced. So why are there so few data showing that VR increases learning compared to other media?

The critical challenge, I believe, is similar to the problem faced by creators of VR entertainment: finding a balance between distraction and narrative in the lesson. Effective teaching requires the instructor to engage in a type of storytelling, to provide context for the facts that are being presented. In some of our early studies, we would narrate the facts and arguments of the lesson while immersing the user in exciting visual events. For example, in a marine science lesson, the student immersed in VR might be looking at a coral reef while the voice of the narrator tells him about how coral changes over time. The problem is, coral looks pretty cool in VR. So cool, that the learner's attention might be so focused on the digital environment that he's not hearing the lesson. While it sounds pretty obvious in retrospect, I believe the key is to separate the engaging experiences—which, according to Dede, motivate learning—from the presentation of learning materials. The problem is, this in some ways defeats the purpose of learning in VR, if the experiences need to be separate from the presentation of lessons.

The answer, I think, is to create experiences that don't require narrative, or the presentation of facts, at all. If we want to fully unlock the potential of VR learning, the lesson should simply emerge from the experience as an active process of discovery. Or a VR experience should alternate between doing and telling, periods of discovery followed by narrative to encapsulate the discoveries. Easy to say, but hard to build.

But let's assume for a moment that VR is no more effective at teaching than video, as measured by controlled experiments in which half the subjects were exposed to video while the other half were exposed to VR. There is still the motivation angle. VR makes learning fun.

In 2016 we set up two fully immersive systems at the Tribeca Film Festival, and craftily disguised a marine science learning field trip as a VR entertainment experience. Jane Rosenthal, a Hollywood producer and the cofounder of the festival, is one of the biggest thinkers in VR entertainment. She and her colleagues set up an arcade for New Yorkers and guests of the festival to experience VR, in a huge floor space with VR demos as far as the eye could see. The festival ran for 12 hours a day, 6 days in a row. We had two booths showing our ocean acidification field trip. All told, we took about 2,000 people through that field trip that week. There was a constant line—sometimes with dozens of people waiting in the queue. They waited for hours to learn about marine science. They even paid for it. I remember thinking at the time that I had never seen people bickering about the wait-time to read a textbook. Of course, much of the excitement was due to the novelty of VR for an audience that had never experienced it. But I'm not even sure if that makes a difference. As the technology gets better and the content gets more sophisticated, will a well-executed VR experience ever seem tired? It hasn't for me in the past two decades. There are no constraints in VR—the only limitation is the imagination. I firmly believe that for people who love to learn, the future is going to be filled with thrilling educational experiences.

But there are other educational advantages to VR beyond its ability to transport us to educationally enriching environments. For the rest of the chapter I want to talk about another intriguing— and potentially sinister—affordance of the medium. That is, the educational potential of analyzing the vast amounts of data that computers can collect on people using VR, providing a durable record of the way they move, speak, and see while inhabiting virtual environments.

There is a classic *New Yorker* cartoon, which has a picture of two dogs surfing the web; the caption is "On the internet, Nobody knows you are a dog." I often tell my students that in VR, we not only know you are a dog, we know what breed you are, the type of collar you are wearing, and what you ate for breakfast. In the history of media use, and for that matter in social science research, there has never been a tool that measures human gestures as accurately, as often, and as unobtrusively as immersive VR. And the data it can collect are intimate and revealing: unlike speech, nonverbal behavior is automatic, a direct pipeline to our mental states, emotions, and identities. We can all watch what we say, but very few of us can consistently regulate our subtle movements and gestures. I've been studying these "digital footprints" for nearly a decade, and along the way I have collected massive amounts of data about the way people move their bodies or direct their gaze. In that time, my lab and others have continued to refine our methods of understanding this information—what Cornell professor Debra

Estrin calls "Small Data Analysis." This work has allowed us to see the behavioral "tells" that predict mistakes on factory floors, or bad driving behavior, or even when someone is intrigued by a product during online shopping. There are all sorts of applications of this technology—some positive, some downright creepy. But in my opinion, one of the best uses of the digital footprints is to assess learning behavior.[7]

Over the course of a semester in school, students will (ideally) have spent many hours in and out of the classroom learning the material of the course. But when it comes to evaluating how much they've learned, there actually isn't much to look at. After months of study, a student's grade is determined by a handful of data points: a midterm, a final, class attendance, and maybe a paper or two. A grade is determined by very few inputs, even though it can dictate whether or not you get into graduate school, get a job, or score a high salary. It is supposed to predict how well a student will later do in the world, and for potential employers it's supposed to say something about how disciplined, conscientious, and hardworking a student is.

A class held in immersive VR—whether it is a short field trip or a longer lecture delivered by a virtual teacher—will generate a massive amount of gesture data, and from these data we can learn a lot about students' engagement and performance. For instance, in one study published in 2014, we used a VR tracking system to gather nonverbal data during one-on-one, student-teacher interactions. Later, we used that data to predict the students' test scores.[8] Results showed that by analyzing the body language of teachers and learners while a class was being taught, we could accurately predict the test score of the students later on. The VR system could know, during the lecture, if a test score from the learner would be high or low. What makes this type of experiment so powerful is the "bottom up" nature of the research. Instead of looking for specific known gestures, like nodding or pointing, we mathematically uncovered subtle movement patterns, many of which would not be noticed by the human eye. One feature of student movement which was predictive of scores was "summed skewness of head and torso" movements. It's very difficult to visualize what this mathematical distribution would look like, but one example of a body movement that might produce this skewness would be occasional nods in a person who otherwise kept their head more or less upright.

Every year I ask students in my Virtual People class a question. Would you rather have your grade based on a few high-pressure hours taking a final exam—the traditional method—or on an analysis of your digital footprint, a continuous measure of learning and engagement formed from the analysis of literally millions of data points spanning hours per week for months at a time? So far, I've found very few takers for the new way. Even though most agree that the digital footprint would be a better way to gauge how good a student is, they vote for the final exam. Despite the higher accuracy of digital footprints, students have gotten quite used to working within the test-taking system. Perhaps they are very comfortable with being great learners only some of the time.

But the ability to detect all movements in real time goes beyond assessment. Virtual teachers can transform and adjust on-the-fly.

My avatar, in a class held entirely in VR, can outperform me as a face-to-face teacher any day. It can pay perfect attention to every student in a class of 200 or more; show my most appropriate actions while concealing any mistake, like losing my composure; and detect the slightest movement, hint of confusion, and improvement in performance of each student simultaneously.

The prevailing wisdom in teaching, as in just about every form of social interaction, is that face-to-face contact is the gold standard, beating all forms of mediated interactions. But my research with avatars and learning argues that a teacher's avatar has powers that just don't exist in physical space.

Virtual reality functions in cycles—the computer figures out what someone is doing, then redraws his or her avatar to show changes based on that behavior. For example, a student in Philadelphia can move his head, look toward the teacher, and raise his hand, and all these actions can be measured by sensing technology. As the student moves, the computer of the teacher in Santa Fe—which already has an avatar with the student's facial features and body shape—receives that information over the Internet and modifies the avatar to make it move in response. Tracking the actions of teacher and students, transmitting them online, and applying them to the respective avatars all occur seamlessly, enabling the participants to feel as if they are in the same virtual room. Each user's computer sends the other machines a stream of information summarizing that user's current state.

However, users can bend reality by altering their streams in real time for strategic purposes. For example, a teacher can choose to have his computer never display an angry expression, but always replace it with a calm face. Or he can screen out distracting student behaviors, like tapping pencils on desks or texting on cellphones.

Research by Benjamin S. Bloom in the 1980s and subsequent studies have demonstrated that students who receive one-on-one instruction learn substantially better than do students in traditional classrooms. Virtual reality makes it possible for one teacher to give one-on-one instruction to many students at the same time—a one-on-100 class, from a nonverbal standpoint, can feel like 100 one-on-ones.

Students in a classroom, like most large groups of people, include a wide range of personality types, among them the usual range of introverts and extroverts. Some students might prefer communication accompanied by nonverbal cues, like gestures and smiles; others may prefer a less-expressive speaker. A number of psychological studies have demonstrated what is called the "chameleon effect": when one person nonverbally mimics another, displaying similar posture and gestures, he maximizes his social influence. Mimickers are seen as more likable and more persuasive than non-mimickers.

In a number of laboratory studies of behaviors including head movements and handshakes in virtual reality, my colleagues and I have demonstrated that if a teacher practices virtual nonverbal mimicry—that is, if she receives the students' nonverbal actions and then transforms her nonverbal behavior to resemble the students' motions—there are three results.

First, the students rarely are conscious of the mimicry.

Second, they nonetheless pay more attention to the teacher: they direct their gaze more at mimicking teachers than they do at teachers who are behaving more normally.

Third, students are influenced more by mimicking teachers— more likely to follow their instructions and to agree with what they say in a lesson.[9]

When I teach a class of 100 students face-to-face, I try to match my nonverbal behavior to that of a single student, and I am forced to devote ample cognitive resources to that effort. But in a virtual classroom, my avatar could seamlessly and automatically create 100 different versions that simultaneously mimic each student. Without my having to pay any attention to my actions, let alone to type commands on a keyboard, my computer could change my gestures and other behaviors to imitate each student's gestures and behavior. In effect, I can psychologically reduce the size of the class.

Historically, one of the most successful uses of virtual reality has been to visualize factors that are impossible to see in the real world. Ivan Sutherland, in his landmark 1965 paper, "The Ultimate Display," makes an early case for this use of VR. Sutherland points out that each of us, in our lifelong experience of the physical world, develops a set of expectations about its properties. From our senses and experience we can make predictions about how physical objects will behave with gravity, or react to each other, or appear from different perspectives. However, these predictions—very useful in our day-to-day physical existence—are misleading when we try to understand more subtle or hidden physics. "We lack," he writes, "corresponding familiarity with the forces on charged particles, forces in non-uniform fields, the effects of non-projective geometric transformations, and high-inertia, low friction motion. A display connected to a digital computer gives us a chance to gain familiarity with concepts not realizable in the physical world. It is a looking glass into a mathematical wonderland."[10]

The final phrase, "a looking glass into a mathematical wonderland," has been a guiding idea for many in the VR community as it has developed since the 1960s.

One of the scholars most associated with using VR to visualize science learning is Andries Van Dam from Brown University, who has spent decades building technology to display "hidden" scientific relationships through computer visualization techniques. Throughout his distinguished career, he has collaborated with experts in medicine, anthropology, geography, and other domains to create tools to enhance learning and facilitate new scientific insights. These simulations can take information in the form of numbers or two-dimensional representations, and turn them into dynamic, inhabitable environments. This type of work allows astronauts to visit the surface of Mars to understand navigation and scale, or biologists to shrink down to the size of a cell to get a new perspective on the structure and function of blood flow. Archeologists get to explore ruins not as they are now—crumbling and only hinting at their past—but as they were, with the structure complete and the artifacts intact.

The field of visualization research history—there are hundreds of studies in this area—is filled with examples of "aha" moments that scientists achieved from getting to walk around their data.

Consider an early example, the ARCHAVE system, which was built to analyze lamp and coin finds in the excavation trenches at the Petra Great Temple site in Jordan. This system provided the ability for archeologists to walk through a VR reconstruction of the site. But more importantly, it allowed visual access to the excavation database—artifacts and data that had been collected over time—enabling scholars to step into VR and examine the data in a life-size representation of the temple. When the computer scientists who built the system brought in archeologists, they learned things from the visualization that would have taken months to learn otherwise. For example, one expert visually examined the presence of lamps and coins, artifacts that were in the database. She discovered a cache of Byzantine lamps in a trench of the western aisle—the grouping only became obvious with the visualization—which indicated important evidence about who might have lived there during the Byzantine occupation.[11]

VR is not going to replace classrooms overnight. Nor should it. What I look forward to seeing, and indeed will continue to actively promote, is a slow, careful, but steady trial-and-error integration of this new and powerful technology into the classroom.

CHAPTER 10

HOW TO BUILD GOOD VR CONTENT

As I write this final chapter, Christmas 2016 has just passed and we're in the middle of Chanukah, which had an unusually late start this year. It's prime gift-giving season, and as many in the industry predicted, millions of people recently unwrapped a Vive, Rift, PlayStation VR, or some other form of virtual reality device gifted from a friend or family member. I imagine many of them, when faced with this unusual contraption, responded as my grandfather did when I gave him his first experience of virtual reality in 2014: "What am I supposed to do with this?" Then, they did what we all do nowadays when we have a question—they asked Google. According to Google Trends, the use of the search term "VR Content" tripled between December 23 and December 26 of 2016. Not coincidentally, so did queries for "VR Porn."[1] It seems people received their goggles, tried out a few stock demos that came with the devices, and then sought out the products of the adult entertainment industry, so often in the vanguard of our developing media technologies—from the VCR to Internet video streaming. In the words of Bill Gates, "content is king," and how VR fares as a consumer technology will depend on how quickly customers find other things they want to do with their devices.

In the late '90s, my initial research wasn't geared toward VR as a consumer product. I worked in a psychology department, and we viewed VR as a tool for a handful of scientists, not a widget that sits next to the TV set. While it sounds strange to say it today, in terms of budget, logistics, and general use, my lab at UCSB likened the VR system to a functional Magnetic Resonance Imaging (fMRI) machine—a ridiculously expensive, bulky technology that needed constant maintenance and could only be administered by trained experts. Because there was no chance our work would filter out to the general public, we were free to use this tool to answer any question we fathomed. Jim Blascovich, my mentor and coauthor of my previous book *Infinite Reality*, designed Vegas-like casinos to test conformity and even tested the effects of experiencing "avatar death." Jack Loomis, another mentor of mine, built surreal rooms with no floors or ceilings to test the limits of the vestibular system. And as you've already read, Skip Rizzo built systems to treat PTSD in soldiers, while Joanne Difide treated survivors of 9/11, and Hunter Hoffman built high-end systems to explore pain reductions in clinical settings. It was pure science, and important work, but never was it conducted with the expectation that laypeople would be opening VR gifts on Christmas morning. Like the fMRI machine, we expected it to be coupled with "adult supervision" for the foreseeable future.

Then I arrived at Stanford, and switched from a psychology department that viewed VR as a tool to understand basic brain science, to a communication department. In communication, we study media use. Hence my thinking evolved, and I began to imagine a world in which avatars and virtual reality were everywhere.

I was on the quest for tenure at Stanford, and decided to bring a new twist, virtual reality, onto one of the pillars of communication, "Media Effects." Media effects asks the simple question—how does media use change people? I imagined a future world as William Gibson or Neil Stephenson described it, with VR being pervasive, and then started studying how VR would throw a wrench into that world. Could politicians use avatars to fix elections? Could VR make advertisements more persuasive? Could changes in the weight of your avatar change the way you eat in the real world? Still, while all this work was happening, I could sleep at night just fine, as VR was still a pipe dream as far as the consumer market was concerned. The technology was supervised, and still confined to those with six-figure budgets and the engineers necessary to keep the systems running.

It was about 2010 when my thinking started to change. One reason might be that I was starting a family (my first child was born in 2011), or that I was witnessing the first wave of VR consumer technology flourish when Microsoft created the Kinect. Perhaps I became influenced by new mentors, for example Jaron Lanier, whose vision of VR was synonymous with a hippie-inspired notion of self-transformation, and Philip Rosedale, whose unbridled enthusiasm for a prosocial world of networked avatars was infectious. Or maybe I had just finally drunk the Silicon Valley Kool-Aid ("making the world a better place"). Regardless, I made a decision. VR was a powerful medium, unlike others that have come before it. It wasn't a media experience—it was an experience—one that occurs "on demand," at the touch of a button. We shouldn't be creating experiences in VR that we wouldn't want to have in the real world.

What, then, are the types of things we could create, and how should we go about doing it? People fly to my lab from all over the world to hear answers to those questions. One of the questions I get asked most often by companies who are trying to get into the VR space is, "What should we do?" My answer, of course, depends on the context. I have had hundreds of these conversations since 2014, when VR began to go mainstream. Here is a set of loose guidelines that have emerged and evolved from those conversations:

1) Ask yourself, does this *need* to be in VR?

VR, like all media, is neither good nor bad—it is a tool. And while I am fascinated and bullish about the amazing experiences that will be shared, the social possibilities it will open up, and the creativity it is going to unleash, I would be remiss not to remind us of the costs of VR.

As I've said elsewhere: presence in VR leads to absence in the physical world. Your mind can't be in two places at once. Using VR carelessly will cause us to step on the dog's tail, walk into walls, and get mugged on subways. Second, the hardware is uncomfortable. Twenty minutes inside most commercial systems available today will leave indentation lines on one's forehead and a bit of grogginess. Third, VR is compelling, and likely addictive—as we've discussed earlier in the book—if the best experiences imaginable can be had at the press of a button.

I've come up with a few rules of thumb for how I think about appropriate VR use cases. First, VR is perfect for things you *couldn't* do in the real world, but not for things you *wouldn't* do in the real world. Flying to the moon like Superman is okay. Participating in virtual mass murder—especially if it is designed to be realistic—is not. This is a big deal. Everything we know about training in VR points to the realization that the medium is incredibly influential on future attitudes and behavior. We don't want to train future terrorists on VR, or desensitize people to violent acts.

Second, don't waste the medium on the mundane. VR experiences should be engaged in mindfully. If five years from now, people are shutting themselves off from the physical world just to read their email, then I have failed in this book. Since we are worried about distraction and addiction, we should save VR for the truly special moments.

The easiest standard to describe is doing the impossible. If it's simply not an option to have a behavior in the real world, then VR is a very safe bet. Time travel is not an option outside of Hollywood, so if you want to go back and meet your great-great-great-grandfather, or feel what it is like to walk around as a cow, or grow a third arm to be more productive in your daily tasks, then you should enter VR.

Using VR to safely experience dangerous behaviors is another good use. As we learned earlier in the book, a primitive version of VR technology was developed to simulate flight in the late 1920s. Why did we build flight simulators? Because mistakes are free in VR, but lives (and planes) are expensive. So, if we can "demote" some of the classic errors to the simulators, then we will be better off. It's time to extend this model of military training to firefighters, nurses, and police officers. I recently gave a tour of my lab to my neighbor, who has been a detective with the Oakland Police force for 20 years. When he experienced the football training simulator described in the first chapter, he instantly thought of riot training. Crowd control is a routine challenge for law enforcement, but it is impossible to build an "obstacle course" to train officers to handle unruly crowds. According to my neighbor, the first time, in full gear, he encountered a group of angry rioters, he faced the most foreign and challenging situation he had ever been a part of as a police officer, even though he was there to keep everyone safe. Imagine how much more in control he and his fellow officers would have felt if they had had dozens of "reps" in virtual practice. I get phone calls and e-mails from police officers fairly regularly. They see this technology as transformative.

Concerns over cost and availability should also guide VR use. Traveling to the top of Mount Kilimanjaro is likely not impossible or dangerous for some of us, but it is a huge investment of time and money. Even those who are physically capable of such a feat are for the most part not blessed with the wealth to do so. Not only will VR afford stunning mountaintop views at a fraction of the expense and exertion, it will also save precious time. I once spent 40 hours—this is a work week—traveling back and forth to give a 45-minute talk in South Africa. If I could have used an avatar to give that talk, it would be like having a 53-week-long year.

This cost consideration certainly applies to medical training, which is full of examples in this domain. Consider how one typically trains surgeons. Cadavers are expensive and rare, and each organ can only get cut once. With VR, there may be cost on the front end to build the simulation, but once it is done, the simulation is like anything digital—you can make a billion copies, send it around the world with a touch of a button—and it lasts forever without degrading.

Some experiences might lead to better behavior, but with bad short-term consequences. When I was growing up, you would hear stories about kids caught smoking cigarettes being made to smoke an entire pack to teach them a lesson—tough love, Baby Boomer style. This might be an effective teachable moment, but this style of education will certainly do damage to the kid's lungs. In VR you can have the best of both worlds. No, you can't simulate the pain of inhaling large amounts of toxic smoke, but you can show the long-term effects of smoking on an avatar, or take a child on a guided tour of damaged lungs. As demonstrated in Chapter 4, you can illustrate the costs of behavior that damages the environment, without actually chopping down real trees. It feels real, and the brain treats it like an experience, but no environmental damage occurs.

The bottom line is, if an experience is not impossible, dangerous, expensive, or counterproductive, then you should seriously consider using a different medium—or even doing it in the real world. Save VR for the special moments.

2) Don't make people sick.

If you do decide to create something in VR, then this should be your primary concern. Good VR feels great. It is fun, engaging, exciting, potentially transformational, and can be an all-around positive experience. That being said, causing people to experience simulator sickness is an absolute deal-breaker. I worry that all it will take is a few very public and visual instances of simulator sickness not only to completely derail a specific VR experience for a given producer, but to actually decelerate the entire VR movement itself.

When I first started working in the field, we had an incident at UC Santa Barbara. A woman in her forties participated in an experiment and got a small case of simulator sickness. Back then, bouts of sickness were fairly common: we were only running at an update rate of 30 frames per second (compared to 90 now) and the latency—the lag between head and body movements and the world updating accordingly—was very high, producing a disconcerting lag as the world always seemed to be a bit behind. The response to these instances—approved thoroughly by the Institutional Review Board at UCSB—was to sit the participant down, and then have her drink some ginger ale and let us know when she felt better. With this particular woman, we followed protocol. After a short time she said she felt fine. We said good-bye to her, and she left.

The next day we got a phone call—the woman drove home, parked at her house, and while walking from her car to her doorstep, got dizzy, fell into a fencepost, and hit her head. It was a really bad day for my colleagues and me. She ended up being just fine, without serious injury, and there were no legal consequences. But it was a reminder: avoid simulator sickness at all costs.

Designers should also be careful not to move the field of view. Let the user move it. I recently was at a trade show and watched one of the largest automobile companies give simulator sickness to one CEO visitor after the next. They put the executives in cars inside the HMD, and drove them around sharp turns, accelerating and decelerating the users, putting the collective vestibular system of the conference through the proverbial ringer.

Why was virtually driving such a strain on the senses? For the past hundreds of thousands of years, when a human has moved, three things have happened. First, your optic flow changes. This is a fancy way to say that if you walk closer to a rock, it gets bigger in your field of view. Second, your vestibular system reacts. Sensitive apparatus, for example in the inner ear, vibrate with movement and provide cues that you are moving. Third, you get proprioceptive cues from the skin and muscles in your body—for example, the bottom of your feet feel pressure as they touch the floor during walking.

Driving in VR wreaks havoc on this system. The drivers at the tradeshow were seeing the road zip by with the proper optic flow, but didn't get the vestibular cues since their body wasn't actually rotating with the car as it whipped around turns. They also didn't get the proprioceptive cues that would cause them to feel pressure from the seat back on their back muscles as the car suddenly accelerated.

This driving anecdote points to a bigger, fundamental issue of movement in VR spaces. People want to explore very large virtual spaces—for example, walk around on the surface of the moon—but very few people have "moonscape-sized" rooms in their homes. Consequently, the challenge is to allow people to explore virtual spaces that are larger than the physical ones that house the hardware and the user. Unfortunately, most solutions are no fun for the human perceptual system, as VR can give the proper optic flow—that is, we can see movement that looks real—but the corresponding vestibular and proprioceptive cues don't match because we aren't actually walking around a very large space.

There have been some fascinating attempts to solve this problem over the past few decades. The best is to have a massive room. A former officemate of mine from UCSB, Dave Waller, started a VR lab at Miami University, Ohio. They gave him a retired gymnasium, which he aptly named the HIVE—Huge Immersive Virtual Environment. While this is likely the best solution for the perceptual system, not many of us have such a luxury of gym-sized VR rooms.

One of the funniest solutions is the human hamster ball. Some early military prototypes popularized this—just like a hamster running in a ball, humans can run in very large spheres that rotate with their footsteps and get the perception of an infinitely large space. Of course, one needs a very large space—including extremely high ceilings—to make this work, as it takes a pretty large sphere not to make the floor seem curved on the feet. This is not likely to be a viable solution for consumer use.

There has been a lot of recent traction on omnidirectional tread-

mills. These are basically treadmills that allow you to walk continuously by having the treads adjust to your walking direction. As the
user walks, the treads automatically pull the user back to the center with treads that move in the opposite direction than the user.
So if you make a right turn in VR, then the treads adjust direction
and speed in order to bring you back to the center of the treadmill.
These systems have improved greatly over the past decade, but still
don't approximate realistic vestibular and proprioceptive cues, as
the walking style needs to be adjusted to fit the device. Also, safety
is a big issue, and most commercial systems are opting to use har-
nesses to ensure someone doesn't fly off the treadmill.

One way to solve the problem is simply to use abstract cues to
move, in the same way videogames use joysticks, mouse clicks, or
arrow buttons. While this is likely the simplest solution, and works
fine for desktop VR systems such as Second Life, it is the most
egregious way to break the coupling of optic flow with the other
two cues. In my lab, if we need to move the user for some reason by
using the mouse, and he sees his field of view lurch forward without
actually moving, more often than not he will shout. Sometimes he
even falls down.

Along these lines, the current solution most VR systems have
adopted is the "Teleport" method. People hop from one location to
another, but the designers do not give any perceptual cues about
the movement itself. By pointing a hand controller to a spot—
similar to a laser pointer in the real world—and clicking a button,
the user can jump to a new location. This sounds jarring, but is
actually surprisingly comfortable compared to the other abstract
methods that render the optic flow of the journey itself. The idea is
that if someone has a room about the size of a typical bedroom, she
can teleport to various locations; then she can walk around natu-

rally in a way that provides proper optic flow, vestibular cues, and proprioceptive cues in the small area. The key is to have the transitions work right during the teleport, so as not to jar the user as she beams from one place to the other.

My favorite technique to solve the room size issue is redirected walking. Imagine a quandary where a user is in a VR simulation that requires him to walk for miles in a straight line. The problem is that he doesn't have a VR space that size. Instead, he has access to a large, square physical room. Let's suppose the user starts on the bottom left corner and physically walks along the left wall of the room. In VR, this works perfectly; he just sees himself walking straight. As he gets to the top left corner, the system starts to subtly shift the orientation of his field of view, and rotates him one degree counter-clockwise with every step he takes toward the corner. Since he is attempting to walk in a straight line in VR, without being aware he is doing so, he physically compensates by rotating his body clockwise, as he continues to take steps in the room. This process continues as he approaches the corner. If you think this through to its logical end, the outcome is a person walking continuously in a circle physically, while perceiving himself walking in a straight line in VR. The key here is to have a room large enough so that the adjustments in orientation can be subtle. This is critical for two reasons—first, you don't want to get someone sick by rotating her field of view drastically, and second, it's a lot cooler if she isn't aware of the breaking of physics.

There are so many amazing things one can do in VR—fly through space, run up and down the pyramids, relive famous moments in history—but there is a golden rule: don't make people vomit. Or even get dizzy. If we embrace this rule early in the development of VR, the community and industry will be better off.

3) Be safe.

Good VR causes people to forget they are in the physical world. I have had 70-year-old men randomly attempt to do a backflip in VR and fall into my arms, famous journalists sprint full speed at physical walls, a Russian businessman almost hit me in the head with a roundhouse kick, and famous football coaches smash their hands against a podium to swat at a sprinting virtual player. Ultimately our lab is safe because we always have a talented and vigilant spotter, a person whose job it is to watch every move of the user and catch or restrain him if necessary. Of course, the "spotter" solution will not work at scale. I like to joke that most commercial packages don't come with me. (I am very proud of my spotting skills.) Instead, they come with guidelines, for example, "Please sit down while playing this game," or scanning systems that often (though not always) warn you about walls. It will only take a few visible and horrifying accidents to nip this VR revolution in the bud. My advice to people doing VR is, however much energy you plan to put into safety, triple it.

One way to bolster safety efforts is to keep VR simulations short. Think about some of your most memorable life experiences. Were they hours long, or minutes? In most forms of storytelling, the mantra is "less is more." In VR, this is especially true. Given that most VR simulations are intense—emotionally engaging, perceptually harrowing, and psychologically compelling—five to ten minutes is often enough.

With all the media attention that virtual reality has been getting in the past few years, it's easy to forget that it is not a new technology. It's not even a recent technology. Speculation about what it can do and how it will change the world has been going on for decades. And yet the truth is, nobody knows what the future holds for VR. The best we can do is try to understand how it works and what it's capable of, and then imagine how these factors can meet the needs and desires of human beings. In this book I have outlined what we know about how VR works as a medium, and while its effects are based on sound research, predicting how a technology matures, or how it will actually be implemented, is always a dicey prospect. With past technologies, designers and experts were rarely on point with their forecasts. Who would have thought—given 5G networks and high-resolution screens—that one of the most popular uses of the powerful data phones we carry around with us would involve sending text messages and tweets—the type of information that a nineteenth-century telegraph could handle? Who would have thought that the most elaborate game interface ever invented, the Microsoft Kinect, which can track the entire body as a game controller, would not supplant a traditional Xbox gaming controller?

If the Internet is any guide for how VR will evolve, most people will not just become VR consumers but VR producers as well, the same way people blog, upload YouTube videos, and tweet. This chapter is called "How to Build Good VR Content," and you'll notice there is a lot more on the "how" than on the "content." That's because, as the technology improves and the creation tools develop, the range of ways people can express themselves in VR, and the applications they can build with it, will be bounded only by their imagination. Some of it will be unsavory. While I wholeheartedly agree with the US Supreme Court's determination that digital simulations are protected free speech, I also believe that just because we have the freedom to build anything we want, doesn't mean that we should. We should strive for more than pure sensationalism or escapism in our entertainment. We can make the world a better place—if you'll forgive me the Silicon Valley cliché—if we respect the unique power of the medium and focus on the pro-social aspects of VR. However it shakes out, it's a very special time to be a part of this technological revolution. The coming years are going to be a wild ride.

ACKNOWLEDGMENTS

First and foremost, I'd like to thank my wife, Janine Zacharia. For those of you who read my last book, *Infinite Reality*, you might notice that this book is more focused on the uses of VR that actually help—people, governments, animals, the environment. My last book outlined what one *could* do in VR; this one focuses on what one *should* do. There is a splash of activism in this book, above and beyond the science. Janine has been instrumental in challenging me to do better—to not just focus on what VR is, but on what it can do. She is one of the reasons I have focused our lab research on climate change, empathy, and the other prosocial applications of VR.

Second, I'd like to thank Jeff Alexander, who helped with every facet of the book—interviews, editing, research, giving VR demos to rappers, talking through ideas, and brainstorming arguments. Without Jeff's brilliance and hard work this book would look very different.

My agent, Will Lippincott, literally made this book happen. I had lost the fire to write a second book, but Will's patience, encouragement, and wisdom are directly responsible for me to take the plunge. Thanks also to my editors. Brendan Curry sculpted the overall arc of the book, and Quynh Do "batted 1000"—every edit she made improved the book.

I have been beyond fortunate to be the primary advisor of ten PhD students—Sun Joo Ahn, Jakki Bailey, Jesse Fox, Fernanda Hererra, Rene Kizilcek, Soo Youn Oh, Kathryn Segovia, Ketaki Shriram, Andrea Stevenson Won, and Nick Yee. In this book I often say that "we" conducted a study. The truth is the PhD students do an overwhelming majority of the thinking and the labor. The lab would have very little to offer without these brilliant minds, and I appreciate their dedication and talent, and their kindness and tolerance when I overstate my contributions to their work. Thanks also to dozens of Stanford undergraduates and masters students who have been essential parts of this research, and to Tobin Asher, Neal Bailenson, Kate Bettinger, Michael Casale, Albert Kashchen-evsky, Shelby Mynhier, and Janine Zacharia for comments on early drafts.

Very few professors are lucky enough to have full-time staff members. Since 2010 I have been completely reliant on my lab managers—first Cody Karutz, then Shawnee Baughman, and now Tobin Asher. Elise Ogle has been an essential project man-ager. Since my lab has been formed at Stanford we have given over 10,000 people VR tours and demonstrations. Professors rarely have time to engage in outreach, whether it is giving a tour or cre-ating a VR film for Jane Rosenthal to premiere at the Tribeca Film Festival, and my amazing staff has made the lab not just a home for science but a place where anyone—whether they are a third-grade class on a field trip or a billionaire CEO scouting an acquisition—can come and learn about VR.

I dedicated this book to Clifford Nass, my faculty mentor. Cliff was a genius, but more than that he was the most caring and spe-cial professor I have ever met. I wouldn't have a job at Stanford without him. His mentorship on my road to tenure cannot be over-stated. Also, some readers may note similarities to the underlying premise of this book—that we should view VR experiences as real ones—as familiar. The *Media Equation*, by Byron Reeves and Clifford Nass, makes the original argument, that we treat media as if it were real. I thank both Cliff and Byron for my intellectual and academic development at Stanford.

There were scores of other mentors. Jim Blascovich taught me social psychology. Andy Beall taught me everything about VR—coding, hardware, big thinking. Jack Loomis taught me about the perceptual system. Jaron Lanier taught me that VR should be about transformation—making us better. I often think I have a truly unique idea, and then find that Jaron had the same idea a few decades prior. Walter Greenleaf and Skip Rizzo taught me about medical and clinical VR. Mel Slater has done more to understand human behavior in VR than just about anyone. Roy Pea and Dan Schwartz taught me about learning and education. Bruce Miller opened my eyes to the world of enterprise, and how to talk to business folks. Mel Blake further honed those skills—without his patience and wisdom, I would never have been able to bring the vision I have for VR, as outlined in this book, to so many decision makers at so many companies. Dirk Smits and Schuyler Cullen helped me begin to fathom the beast that is Silicon Valley (I still have miles to go). Carole Bellis provided not only sage legal counsel along the way but managed to make it fun.

A special thanks to Derek Belch, who will have a place in VR history as one who mainstreamed VR, changing the way people train. Most of my colleagues work hard, but as far as I can tell, nobody works harder than Derek.

Research is expensive. I have had the pleasure of working with funders who were not only generous but typically wise and helpful as well. Thank you to: Brown Institute, Cisco, Coral Reef Alliance, Dai Nippon Printing Co. Ltd., Defense Advanced Research Projects Agency (DARPA), Google, Gordon and Betty Moore Foundation, HTC Vive, Konica Minolta, Media-X, Microsoft, National Institutes of Health, National Science Foundation, NEC Corporation, Oculus, Office of Naval Research (ONR), OMRON Corporation, Robert Wood Johnson Foundation, Stanford Center on Longevity, Stanford Office of Technology Licensing, Stanford Vice Provost for Undergraduate Education, Stanford Woods Institute for the Environment, Time-sharing Experiments for the Social Sciences, US Department of Energy, and Worldviz.

Finally I'd like to thank the people who are ultimately responsible for my success—Eleanor and Jim, and Neal and Myrna. A wise woman once told me that if you can choose one thing about your life, choose your parents. I wouldn't change a thing, and that also goes for my new family, Richard and Debra Zacharia.

And of course the lights of my life, Anna and Edie.

NOTES

INTRODUCTION

1. "Oculus," *cdixon blog*, March 25, 2014, http://cdixon.org/2014/03/25/oculus/.
2. "Insanely Virtual," *The Economist,* October 15, 2016, http://www.economist .com/news/business/21708715-china-leads-world-adoption-virtual-reali ty-insanely-virtual.

1. PRACTICE MADE PERFECT

1. Bruce Feldman, "I Was Blown Away: Welcome to Football's Quarterback Revolution," *FoxSports*, March 11, 2015, http://www.foxsports.com/ college-football/story/stanford-cardinal-nfl-virtual-reality-qb-training-031115.
2. Author interview with Carson Palmer, June 9, 2016.
3. Peter King, "A Quarterback and His Game Plan, Part I: Five Days to Learn 171 Plays," *MMQB*, Wednesday, November 18, 2015, http://mmqb.si .com/mmqb/2015/11/17/nfl-carson-palmer-arizona-cardinals-inside-game-plan; Peter King, "A Quarterback and His Game Plan, Part II: Virtual Reality Meets Reality," *MMQB*, Thursday, November 19, 2015, http:// mmqb.si.com/2015/11/18/nfl-carson-palmer-arizona-cardinals-inside-game-plan-part-ii-cleveland-browns.

4. Josh Weinfuss, "Cardinals' use of virtual reality technology yields record season," *ESPN*, January 13, 2016, http://www.espn.com/blog/nflnation/post/_/id/195755/cardinals-use-of-virtual-reality-technology-yields-record-season.

5. M. Lombard and T. Ditton, "At the Heart of it All: The Concept of Presence," *Journal of Computer-Mediated Communication* 3, no. 2 (1997).

6. James J. Cummings and Jeremy N. Bailenson, "How Immersive Is Enough? A Meta-Analysis of the Effect of Immersive Technology on User Presence," *Media Psychology* 19 (2016): 1–38.

7. "Link, Edwin Albert," *The National Aviation Hall of Fame*, http://www.nationalaviation.org/our-enshrinees/link-edwin/.

8. National Academy of Engineering, Memorial Tributes: *National Academy of Engineering, Volume 2* (Washington, DC: National Academy Press, 1984), 174.

9. James L. Neibaur, *The Fall of Buster Keaton: His Films for MGM, Educational Pictures, and Columbia* (Lanham, MD: Scarecrow Press, 2010), 79.

10. Jeremy Bailenson et al., "The Effect of Interactivity on Learning Physical Actions in Virtual Reality," *Media Psychology* 11 (2008): 354–76.

11. Feldman, "I Was Blown Away."

12. Daniel Brown, "Virtual Reality for QBs: Stanford Football at the Forefront," *Mercury News*, September 9, 2015, http://www.mercurynews.com/49ers/ci_28784441/virtual-reality-qbs-stanford-football-at-forefront.

13. King, "A Quarterback and His Game Plan, Part I"; King, "A Quarterback and His Game Plan, Part II."

14. K. Anders Ericksson and Robert Pool, *Peak: Secrets from the New Science of Expertise* (New York: Houghton Mifflin Harcourt, 2016), 64.

15. B. Calvo-Merino, D. E. Glaser, J. Grèzes, R. E. Passingham, and P. Haggard, "Action Observation and Acquired Motor Skills: An fMRI Study with Expert Dancers," *Cerebral Cortex* 15, no. 8 (2005): 1243–49.

16. Sian L. Beilock, et al. "Sports Experience Changes the Neural Processing of Action Language," *The National Academy of Sciences* 105 (2008): 13269–73.

17. http://www.independent.co.uk/environment/global-warming-data-centres-to-consume-three-times-as-much-energy-in-next-decade-experts-warn-a6830086.html.

18. http://www.businessinsider.com/walmart-using-virtual-reality-employee-training-2017-6.

2. YOU ARE WHAT YOU EAT

1. Stanley Milgram, "Behavioral Study of Obedience," *Journal of Abnormal and Social Psychology* 67, no. 4 (1963): 371–78.

2. Mel Slater et al., "A Virtual Reprise of the Stanley Milgram Obedience Experiments," *PLoS One* 1 (2006): e39.

3. Ibid.

4. K. Y. Segovia, J. N. Bailenson, and B. Monin, "Morality in tele-immersive environments," Proceedings of the International Conference on Immersive Telecommunications (IMMERSCOM), May 27–29, Berkeley, CA.

5. C. B. Zhong and K. Liljenquist, "Washing away your sins: Threatened morality and physical cleansing," *Science* 313, no. 5792 (2006): 1451.

6. T. I. Brown, V. A. Carr, K. F. LaRocque, S. E. Favila, A. M. Gordon, B. Bowles, J. N. Bailenson, and A. D. Wagner, "Prospective representation of navigational goals in the human hippocampus," *Science* 352 (2016): 1323.

7. Stuart Wolpert, "Brain's Reaction to Virtual Reality Should Prompt Further Study Suggests New Research by UCLA Neuroscientists," *UCLA Newsroom*, November 24, 2014, http://newsroom.ucla.edu/releases/brains-reaction-to-virtual-reality-should-prompt-further-study-suggests-new-research-by-ucla-neuroscientists.

8. Zahra M. Aghajan, Lavanya Acharya, Jason J. Moore, Jesse D. Cushman, Cliff Vuong, and Mayank R. Mehta, "Impaired Spatial Selectivity and Intact Phase Precession in Two-Dimensional Virtual Reality," *Nature Neuroscience* 18 (2015): 121–28.

9. Oliver Baumann and Jason B. Mattingley, "Dissociable Representations of Environmental Size and Complexity in the Human Hippocampus," *Journal of Neuroscience* 33, no. 25 (2013): 10526–33.

10. Albert Bandura et al., "Transmission of Aggression Through Imitation of Aggressive Models," *Journal of Abnormal and Social Psychology* 63 (1961): 575–82.

11. Andreas Olsson and Elizabeth A. Phelps, "Learning Fears by Observing Others: The Neural Systems of Social Fear Transmission," *Nature Neuroscience* 10 (2007): 1095–1102.

12. Michael Rundle, "Death and Violence 'Too intense' in VR, game developers admit," WIRED UK, October 28, 2015, http://www.wired.co.uk/article/virtual-reality-death-violence.

13. Joseph Delgado, "Virtual reality GTA: V with hand tracking for weapons," *veryjos*, February 18, 2016, http://rly.sexy/virtual-reality-gta-v-with-hand-tracking-for-weapons/.

14. Craig A. Anderson, "An Update on the Effects of Playing Violent Video-games," *Journal of Adolescence* 27 (2004): 113–22.

15. Jeff Grabmeier, "Immersed in Violence: How 3-D Gaming Affects Videogame Players," *Ohio State University*, October 19, 2014, https://news.osu.edu/news/2014/10/19/%E2%80%8Bimmersed-in-violence-how-3-d-gaming-affects-video-game-players/.

16. Hanneke Polman, Bram Orobio de Castro, and Marcel A. G. van Aken, "Experimental study of the differential effects of playing versus watching violent videogames on children's aggressive behavior," *Aggressive Behavior* 34 (2008): 256–64.

17. S. L. Beilock, I. M. Lyons, A. Mattarella–Micke, H. C. Nusbaum, and S. L. Small, "Sports experience changes the neural processing of action language," *Proceedings of the National Academy of Sciences of the United States of America*, September 2, 2008, https://wwww.ncbi.mln.nih.gov/pmc/articles/PMC2527992/.

18. Helen Pidd, "Anders Breivik 'trained' for shooting attacks by playing Call of Duty," *Guardian*, April 19, 2012, http://www.theguardian.com/world/2012/apr/19/anders-breivik-call-of-duty.

19. Jodi L. Whitaker and Brad J. Bushman, " 'Boom, Headshot!' Effect of Video-game Play and Controller Type on Firing Aim and Accuracy," *Communication Research* 7 (2012) 879–89.

20. William Gibson, *Neuromancer* (New York: Ace Books, 1984), 6.

21. Sherry Turkle, *Alone Together* (New York: Basic Books, 2011).

22. Frank Steinicke and Ger Bruder, "A Self-Experimentation about Long-Term Use of Fully-Immersive Technology," https://basilic.informatik.uni-hamburg.de/Publications/2014/SB14/sui14.pdf.

23. Eyal Ophir, Clifford Nass, and Anthony D. Wagner, "Cognitive control in media multitaskers," *PNAS* 106 (2009): 15583–87.

24. Kathryn Y. Segovia and Jeremy N. Bailenson, "Virtually True: Children's Acquisition of False Memories in Virtual Reality," *Media Psychology* 12 (2009): 371–93.

25. J. O. Bailey, Jeremy N. Bailenson, J. Obradović, and N. R. Aguiar, "Immersive virtual reality influences children's inhibitory control and social behavior,"

paper presented at the International Communication 67th Annual Conference, San Diego, CA.

26. Matthew B. Crawford, *The World Beyond Your Head: On Becoming an Individual in an Age of Distraction* (New York: Farrar, Straus, and Giroux, 2015), 86.

3. WALKING IN THE SHOES OF ANOTHER

1. Gabo Arora and Chris Milk, *Clouds Over Sidra* (Within, 2015), 360 Video, 8:35, http://with.in/watch/clouds-over-sidra/.

2. Ibid.

3. Chris Milk, "How virtual reality can create the ultimate empathy machine," filmed March 2015, TED video, 10:25, https://www.ted.com/talks/chris_milk_how_virtual_reality_can_create_the_ultimate_empathy_machine #t-54386.

4. Ibid.

5. John Gaudiosi, "UN Uses Virtual Reality to Raise Awareness and Money," *Fortune*, April 18, 2016, http://fortune.com/2016/04/18/un-uses-virtual-reality-to-raise-awareness-and-money/.

6. See Steven Pinker's *The Better Angels of Our Nature* (New York: Viking, 2011) and Peter Singer's *The Expanding Circle* (Princeton: Princeton University Press, 2011).

7. J. Zaki, "Empathy: A Motivated Account," *Psychological Bulletin* 140, no. 6 (2014): 1608–47.

8. Susanne Babbel, "Compassion Fatigue: Bodily symptoms of empathy," *Psychology Today*, July 4, 2012, https://www.psychologytoday.com/blog/somatic-psychology/201207/compassion-fatigue.

9. Mark H. Davis, "A multidimensional approach to individual differences in empathy," *JSAS Catalog of Selected Documents in Psychology* 10 (1980): 85, http://fetzer.org/sites/default/files/images/stories/pdf/selfmeasures/EMPATHY-InterpersonalReactivityIndex.pdf.

10. Mark H. Davis, "Effect of Perspective Taking on the Cognitive Representation of Persons: A Merging of Self and Other," *Journal of Personality and Social Psychology* 70, no. 4 (1996): 713–26.

11. Adam D. Galinsky and Gordon B. Moskowitz, "Perspective-taking: Decreasing stereotype expression, stereotype accessibility, and in-group favoritism," *Journal of Personality and Social Psychology* 78 (2000): 708–24.

12. Matthew Botvinick and Jonathan Cohen, "Rubber Hands 'Feel' Touch That Eyes See," *Nature* 391, no. 756 (1998).

13. Mel Slater and Maria V. Sanchez-Vives, "Enhancing Our Lives with Immersive Virtual Reality," *Frontiers in Robotics and AI*, December 19, 2016, http://journal.frontiersin.org/article/10.3389/frobt.2016.00074/full.

14. N. Yee and J. N. Bailenson, "Walk a Mile in Digital Shoes: The Impact of Embodied Perspective-taking on the Reduction of Negative Stereotyping in Immersive Virtual Environments," *Proceedings of Presence 2006: The 9th Annual International Workshop on Presence,* August 24–26, 2006.

15. Ibid.

16. Victoria Groom, Jeremy N. Bailenson, and Clifford Nass, "The influence of racial embodiment on racial bias in immersive virtual environments," *Social Influence* 4 (2009): 1–18.

17. Ibid.

18. Tabitha C. Peck et al., "Putting yourself in the skin of a black avatar reduces implicit racial bias," *Consciousness and Cognition* 22 (2013): 779–87.

19. Sun Joo (Grace) Ahn, Amanda Minh Tran Le, and Jeremy Bailenson, "The Effect of Embodied Experiences on Self-Other Merging, Attitude, and Helping Behavior," *Media Psychology* 16 (2013): 7–38.

20. Ibid.

21. Arielle Michal Silverman, "The Perils of Playing Blind: Problems with Blindness Simulation and a Better Way to Teach about Blindness," *Journal of Blindness Innovation and Research* 5 (2015).

22. Ahn, Le, and Bailenson, "The Effect of Embodied Experiences," *Media Psychology* 16 (2013): 7–38.

23. Kipling D. Williams and Blair Jarvis, "Cyberball: A program for use in research on interpersonal ostracism and acceptance," *Behavior Research Methods* 38 (2006): 174–80.

24. Soo Youn Oh, Jeremy Bailenson, E. Weisz, and J. Zaki, "Virtually Old: Embodied Perspective Taking and the Reduction of Ageism Under Threat," *Computers in Human Behavior* 60 (2016): 398–410.

25. J. Zaki, "Empathy: A Motivated Account," *Psychological Bulletin* 140, no. 6 (2014): 1608–47.

26. Frank Dobbin and Alexander Kalev, "The Origins and Effects of Corporate Diversity Programs," in *The Oxford Handbook of Diversity and Work*, ed. Peter E. Nathan (New York: Oxford University Press, 2013), 253–81.

27. Sun Joo (Grace) Ahn et al., "Experiencing Nature: Embodying Animals in Immersive Virtual Environments Increases Inclusion of Nature in Self and Involvement with Nature," *Journal of Computer-Mediated Communication* (2016).

28. Caroline J. Falconer et al., "Embodying self-compassion within virtual reality and its effects on patients with depression," *British Journal of Psychiatry* 2 (2016): 74–80.

29. Ibid.

4. WORLDVIEW

1. *Overview,* documentary directed by Guy Reid, 2012; "What are the Noetic Sciences?," *Institute of Noetic Sciences,* http://www.noetic.org/about/what-are-noetic-sciences.

2. In fact, a start-up VR company called SpaceVR has launched a satellite with a VR camera into low Earth orbit and will allow subscribers to gaze at real-time images of the Earth from space.

3. Leslie Kaufman, "Mr. Whipple Left it Out: Soft is Rough on Forests," *New York Times,* February 25, 2009, http://www.nytimes.com/2009/02/26/science/earth/26charmin.html.

4. Mark Cleveland, Maria Kalamas, and Michel Laroche, "Shades of green: linking environmental locus of control and pro-environmental behaviors," *Journal of Consumer Marketing* 29, no. 5 (May 2012): 293–305, 22 (2005): 198–212.

5. S. J. Ahn, J. N. Bailenson, and D. Park, "Short and Long-term Effects of Embodied Experiences in Immersive Virtual Environments on Environmental Locus of Control and Behavior," *Computers in Human Behavior* 39 (2014): 235–45.

6. S. J. Ahn, J. Fox, K. R. Dale, and J. A. Avant, "Framing Virtual Experiences: Effects on Environmental Efficacy and Behavior Over Time," *Communication Research* 42, no. 6 (2015): 839–63.

7. J. O. Bailey, J. N. Bailenson, J. Flora, K. C. Armel, D. Voelker, and B. Reeves, "The impact of vivid and personal messages on reducing energy consumption related to hot water use," *Environment and Behavior* 47, no. 5 (2015): 570–92.

8. "A History of the NOAA," *NOAA History,* http://www.history.noaa.gov/legacy/noaahistory_2.html, last modified June 8, 2006.

9. Intergovernmental Panel on Climate Change, http://www.ipcc.ch/.

10. Remarks from Woods Institute Speech: "Increasingly common experiences with extreme climate-related events such as the Colorado wildfires, a record warm spring, and preseason hurricanes have convinced many Americans climate change is a reality."

11. Daniel Grossman, "UN: Oceans are 30 percent more acidic than before fossil fuels," *National Geographic*, December 15, 2009, http://voices.national geographic.com/2009/12/15/acidification/.

12. Interview with the BBC.

13. Alan Sipress, "Where Real Money Meets Virtual Reality, the Jury Is Still Out," *Washington Post*, December 26, 2006.

5. TIME MACHINES FOR TRAUMA

1. Anemona Hartocollis, "10 Years and a Diagnosis Later, 9/11 Demons Haunt Thousands," *New York Times*, August 9, 2011.

2. "At the time of the WTC attacks, expert treatment guidelines for PTSD, which were published for the first time in 1999, recommended that CBT with imaginal exposure should be the first-line therapy for PTSD." JoAnn Difede et al., "Virtual Reality Exposure Therapy for the Treatment of Posttraumatic Stress Disorder Following September 11, 2001," *Journal of Clinical Psychiatry* 68 (2007): 1639–47.

3. Yael Kohen, "Firefighter in Distress," *New York Magazine*, 2005, http://nymag.com/nymetro/health/bestdoctors/2005/11961/.

4. Interview with JoAnn Defide.

5. JoAnn Defide and Hunter Hoffman, "Virtual Reality Exposure Therapy for World Trade Center Post-traumatic Stress Disorder: A Case Report," *CyberPsychology & Behavior* 5, no. 6 (2002): 529–35.

6. Ibid.

7. Author interview with JoAnn Difede.

8. JoAnn Defide et al., "Virtual Reality Exposure Therapy for the Treatment of Posttraumatic Stress Disorder Following September 11, 2001," *Journal of Clinical Psychiatry* 11 (2007): 1639–47.

6. ABSENCE MAKES THE PAIN GROW FAINTER

1. "Lower Back Pain Fact Sheet," *National Institute of Neurological Disorders*

and Stroke, http://www.ninds.nih.gov/disorders/backpain/detail_backpain .htm, last modified August 12, 2016.

2. Nora D. Volkow, "America's Addiction to Opioids: Heroin and Prescription Drug Abuse," paper presented at the Senate Caucus on International Narcotics Control, Washington, DC, May 14, 2014, https://www.drugabuse.gov/ about-nida/legislative-activities/testimony-to-congress/2016/americas-addiction-to-opioids-heroin-prescription-drug-abuse.

3. Dan Nolan and Chris Amico, "How Bad is the Opioid Epidemic?" *Frontline*, February 23, 2016, http://www.pbs.org/wgbh/frontline/article/how-bad-is-the-opioid-epidemic/.

4. Join Together Staff, "Heroin Use Rises as Prescription Painkillers Become Harder to Abuse," *Drug-Free*, June 7, 2012, http://www.drugfree.org/news-service/heroin-use-rises-as-prescription-painkillers-become-harder-to-abuse/.

5. Tracie White, "Surgeries found to increase risk of chronic opioid use," *Stanford Medicine News Center*, July 11, 2016, https://med.stanford.edu/ news/all-news/2016/07/surgery-found-to-increase-risk-of-chronic-opioid-use.html.

6. "Virtual Reality Pain Reduction," HITLab, https://www.hitl.washington.edu/ projects/vrpain/.

7. "VR Therapy for Spider Phobia," HITLab, https://www.hitl.washington.edu/ projects/exposure/.

8. Hunter G. Hoffman et al., "Modulation of thermal pain–related brain activity with virtual reality: evidence from fMRI," *Neuroreport* 15 (2004): 1245–48.

9. Ibid.

10. Yuko S. Schmitt et al., "A Randomized, Controlled Trial of Immersive Virtual Reality Analgesia during Physical Therapy for Pediatric Burn Injuries," *Burns* 37 (2011): 61–68.

11. Ibid.

12. Mark D. Wiederhold, Kenneth Gao, and Brenda K. Wiederhold, "Clinical Use of Virtual Reality Distraction System to Reduce Anxiety and Pain in Dental Procedures," *Cyberpsychology, Behavior, and Social Networking* 17 (2014): 359–65.

13. Susan M. Schneider and Linda E. Hood, "Virtual Reality: A Distraction Intervention for Chemotherapy," *Oncology Nursing Forum* 34 (2007): 39–46.

14. Tanya Lewis, "Virtual Reality Treatment Relieves Amputee's Phantom Pain," *Live Science*, February 25, 2014, http://www.livescience.com/43665-virtual-reality-treatment-for-phantom-limb-pain.html.

15. J. Foell et al., "Mirror therapy for phantom limb pain: Brain changes and the role of body representation," *European Journal of Pain* 18 (2014): 729–39.

16. A. S. Won, J. N. Bailenson, and J. Lanier, "Homuncular Flexibility: The Human Ability to Inhabit Nonhuman Avatars," *Emerging Trends in the Social and Behavioral Sciences: An Interdisciplinary, Searchable, and Linkable Resource* (Hoboken: John Wiley and Sons, 2015), 1–16.

17. A. S. Won, Jeremy Bailenson, J. D. Lee, and Jaron Lanier, "Homuncular Flexibilty in Virtual Reality," *Journal of Computer-Mediated Communication* 20 (2015): 241–59.

18. A. S. Won, C. A. Tataru, C. A. Cojocaru, E. J. Krane, J. N. Bailenson, S. Niswonger, and B. Golianu, "Two Virtual Reality Pilot Studies for the Treatment of Pediatric CRPS," *Pain Medicine* 16, no. 8 (2015): 1644–47.

7. BRINGING SOCIAL BACK TO THE NETWORK

1. Elisabeth Rosenthal, "Toward Sustainable Travel: Breaking the Flying Addiction," *environment360*, May 24, 2010, http://e360.yale.edu/feature/toward_sustainable_travel/2280/.

2. John Bourdreau, "Airlines still pamper a secret elite," *Mercury News*, July 31, 2011, http://www.mercurynews.com/2011/07/31/airlines-still-pamper-a-secret-elite/.

3. Ashley Halsey III, "Traffic Deaths Soar Past 40,000 for the First Time in a Decade," *Washington Post*, February 15, 2017.

4. Christopher Ingraham, "Road rage is getting uglier, angrier and a lot more deadly," *Washington Post*, February 18, 2015, https://www.washingtonpost.com/news/wonk/wp/2015/02/18/road-rage-is-getting-uglier-angrier-and-a-lot-more-deadly/.

5. "UN projects world population to reach 8.5 billion by 2030, driven by growth in developing countries," *UN News Centre*, July 29, 2015, http://www.un.org/apps/news/story.asp?NewsID=51526#.V9DVUJOUo08.

6. Michael Abrash, "Welcome to the Virtual Age," *Oculus Blog*, March 31, 2016, https://www.oculus.com/blog/welcome-to-the-virtual-age.

7. W. S. Condon and W. D. Ogston, "A Segmentation of Behavior," *Journal of Psychiatric Research* 5 (1967): 221–35.

8. Adam Kendon, "Movement coordination in social interaction: Some examples described," *Acta Psychologica* 32 (1970): 101–25.

9. Adam Kendon, *Conducting Interaction: Patterns of Behavior in Focused Encounters* (Cambridge: Cambridge University Press, 1990), 114.

10. Clair O'Malley et al., "Comparison of face-to-face and video mediated interaction," *Interacting with Computers* 8 (1996): 177–92.

11. Marianne LaFrance, "Nonverbal synchrony and rapport: Analysis by the cross-lag panel technique," *Social Psychology Quarterly* 42 (1979): 66–70.

12. Andrea Stevenson Won et al., "Automatically Detected Nonverbal Behavior Predicts Creativity in Collaborating Dyads," *Journal of Nonverbal Behavior* 38 (2014): 389–408.

13. Scott S. Wiltermuth and Chip Heath, "Synchrony and Cooperation," *Psychology Science* 20 (2009): 1–5.

14. Philip Rosedale, "Life in Second Life," TED Talk, December 2008, https://www.ted.com/talks/the_inspiration_of_second_life/transcript?language=en.

15. "Just How Big is Second Life?—The Answer Might Surprise You [Video Infographic]," YouTube video, 1:52, posted by "Luca Grabacr," November 3, 2015, https://www.youtube.com/watch?v=55tZbq8yMYM.

16. Dean Takahashi, "Second Life pioneer Philip Rosedale shows off virtual toy room in High Fidelity," *Venture Beat*, October 28, 2015, http://venturebeat.com/2015/10/28/virtual-world-pioneer-philip-rosedale-shows-off-virtual-toy-room-in-high-fidelity/.

17. Ibid.

18. J. H. Janssen, J. N. Bailenson, W. A. IJsselsteijn, and J. H. D. M. Westerink, "Intimate heartbeats: Opportunities for affective communication technology," *IEEE Transactions on Affective Computing* 1, no. 2 (2010): 72–80.

19. A. Haans and A. I. Wijnand, "The Virtual Midas Touch: Helping Behavior After a Mediated Social Touch," *IEEE Transactions on Haptics* 2, no. 3 (2009): 136–40.

20. Tanya L. Chartrand and John A. Bargh, "The Chameleon Effect: The Perception-Behavior Link and Social Interaction," *Journal of Personality and Social Psychology* 76, no. 6 (1999): 893–910.

21. David Foster Wallace, *Infinite Jest* (Boston: Little, Brown, 1996), 146–49.

22. S. Y. Oh, J. N. Bailenson, Nicole Kramer, and Benjamin Li, "Let the Avatar Brighten Your Smile: Effects of Enhancing Facial Expressions in Virtual Environments," *PLoS One* (2016).

8. STORIES IN THE ROUND

1. Susan Sontag, *Regarding the Pain of Others* (New York: Farrar, Straus and Giroux, 2003), 54.

2. Jon Peddie, Kurt Akeley, Paul Debevec, Erik Fonseka, Maichael Mangan, and Michael Raphael, "A Vision for Computer Vision: Emerging Technologies," July 2016 SIGGRAPH Panel, http://dl.acm.org/citation.cfm?id=2933233.

3. Zeke Miller, "Romney Campaign Exaggerates Size of Nevada Event with Altered Image," *Buzzfeed*, October 26, 2012, https://www.buzzfeed.com/zekejmiller/romney-campaign-appears-to-exaggerate-size-of-neva.

4. Hillary Grigonis, "Lytro Re-Creates the Moon Landing to Demonstrate Just What Light-field VR Can Do," *Digital Trends*, August 31, 2016, http://www.digitaltrends.com/virtual-reality/lytro-immerge-preview-video-released/.

5. "One Dark Night," *Emblematic*, https://emblematicgroup.squarespace.com/#/one-dark-night/.

6. Adi Robertson, "Virtual reality pioneer Nonny de la Peña charts the future of VR journalism," *The Verge*, January 25, 2016, http://www.theverge.com/2016/1/25/10826384/sundance-2016-nonny-de-la-pena-virtual-reality-interview.

7. The history of how cinematic storytelling evolved from the silent film era through the early talkies is told in the documentary *Visions of Light* (Kino International, 1992), directed by Arnold Glassman, Todd McCarthy, and Stuart Samuels.

9. REVERSE FIELD TRIPS

1. "joan ganz cooney," *Sesame Workshop*, http://www.sesameworkshop.org/about-us/leadership-team/joan-ganz-cooney/.

2. Keith W. Mielke, "A Review of Research on the Educational and Social Impact of *Sesame Street*," in *G Is for Growing: Thirty Years of Research on Children*

and Sesame Street, ed. Shalom M. Fisch and Rosemarie T. Truglio (Mahwah, NJ: Lawrence Erlbaum Associates, 2001), 83.

3. Daniel L. Schwartz and John D. Bransford, "A Time for Telling," *Cognition and Instruction* 16 (1998): 475–522.

4. Chris Dede, "Immersive Interfaces for Engagement and Learning," *Science* 323 (2009): 66–69.

5. S. J. Metcalf, J. Clarke, and C. Dede, "Virtual Worlds for Education: River City and EcoMUVE," *Media in Transition International Conference,* MIT, April 24–26, 2009.

6. Roxana Moreno and Richard E. Mayer, "Learning Science in Virtual Reality Multimedia Environments: Role and Methods and Media," *Journal of Educational Psychology* 94, no. 3 (September 2002): 598–610.

7. "the small data lab @CornellTech," http://smalldata.io/.

8. Andrea Stevenson Won, Jeremy N. Bailenson, and Joris H. Jannsen, "Automatic Detection of Nonverbal Behavior Predicts Learning In Dyadic Interactions," *IEEE Transactions On Affective Computing* 5 (2014): 112–25.

9. J. N. Bailenson, N. Yee, J. Blascovic, and R. E. Guadagno, "Transformed Social Interaction in Mediated Interpersonal Communications," from *Mediated Interpersonal Communications* (New York, Routledge, 2008), 75–99.

10. Ivan E. Sutherland, "The Ultimate Display," in *Proceedings of the IFIP Congress*, ed. Wayne A. Kalenich (London: Macmillan, 1965), 506–8.

11. Andries Van Dam, Andrew S. Forsberg, David H. Laidlaw, Joseph J. LaViola Jr., and Rosemary M. Simpson, "Immersive VR for Scientific Visualization: A Progress Report," *IEEE Computer Graphics and Applications* 20, no. 6 (2000): 26–52.

10. HOW TO BUILD GOOD VR CONTENT

1. Google Trends, https://www.google.com/trends/explore?date=today%203-m&q=vr%20porn.

INDEX